MW01236112

FIRST SEMESTER ORGANIC CHEMISTRY REACTIONS:
EVERYTHING IN ONE PLACE

RHETT C. SMITH, PH.D.

Ideal Quill Publishing Group
idealquill.com

Printed in the United States of America

10 9 8 7 6 5 4 3 2 1

ISBN 978-1463663391

First Semester Organic Chemistry Reactions: Everything in One Place

by Rhett C. Smith, Ph.D.

Contents

Preface – How to Use this Book

Much of what you learn in a first semester introductory organic chemistry class relates to reactions of functional groups. You will explore the reactant / reagent combinations, reaction conditions, stereochemical outcomes and examine reasonable reaction mechanisms (so-called "arrow-pushing") for the transformation of starting materials to products. Although most text books contain a fair deal of information on each of the reactions that you are expected to learn, these critical take-home points are often scattered over a large number of pages amongst a lot of additional text. This concise book, based on my notes from when I took organic chemistry, is designed to provide you with key information for the most-taught key reactions of a first semester organic chemistry course with Everything in One Place.

In addition to simply bringing together key concepts into a compact format, this book also breaks down many of the complicated, multistep arrow-pushing mechanisms into a series of simple steps. In fact, a large portion of reactions (excluding radical and pericyclic reactions) can be broken down into combinations of only six basic mechanistic steps. These six steps are introduced in Section I. The remaining sections are dedicated to outlining and providing practical examples of key reactions.

Following each new reaction, you will find a self-test for that reaction. You will derive the greatest benefit if you take each self-test right after you study each reaction carefully. You should then immediately check your answers (provided on the page following each self-test) so that you get rapid feedback on your level of understanding. It would be valuable to keep a list of reactions that are covered in your class so that you can retake the self tests in the time leading up to each exam /quiz so that you can solidify your long-term understanding and maximize your performance. By the time you finish this book, I hope that you have Everything in One Place – in your mind.

– Rhett C. Smith, Ph.D.
Gainesville, FL
August 2011

I. Basic Mechanistic Steps in Organic Chemistry

The six basic steps of many organic reactions in introductory organic classes are as follows:

1. coordination / heterolysis
2. carbocation rearrangement
3. electrophilic addition / electrophilic elimination
4. E2 reaction
5. S_N2 reaction
6. nucleophilic addition / nucleophilic elimination

These six basic mechanistic steps, in one direction or the other, comprise a large portion of the mechanistic steps that you will observe in the course of learning arrow-pushing mechanisms in organic chemistry. Some reactions that will not employ these common basic steps, but which you will see in an introductory organic chemistry class, include radical reactions and pericyclic reactions.

There are, of course, a handful of reactions that have complicated or poorly understood mechanisms, and some concerted steps (steps of a mechanism that happen all at once) that are a combination of two or more of these basic steps. Where these special examples occur, a special note will be made to point it out. You should find, however, that knowing and looking for these basic steps will be a useful tool in assessing or predicting reaction mechanisms, and recognizing recurring themes as you learn an increasing number of reactions. The rest of this section is devoted to specifying what to look for in identifying each of these basic mechanistic steps, as well as to providing important practical examples and variations that you may encounter.

1. Coordination / Heterolysis (Lewis acid – Lewis base type reaction)

Notice that the donor (Lewis base) can be neutral or anionic, and the acceptor (Lewis acid) can be cationic or neutral. The donor can even be a pair of electrons from a pi bond as in Example 5.

a) General reaction:

$$B^{\ominus} \;\; E^{+} \underset{\text{heterolysis}}{\overset{\text{coordination}}{\rightleftharpoons}} B{-}E$$

b) Example 1:

$$HO^{\ominus} \;\; H^{+} \underset{\text{heterolysis}}{\overset{\text{coordination}}{\rightleftharpoons}} H{-}O{-}H$$

c) Example 2:

$$H_3N{:} \;\; H^{+} \underset{\text{heterolysis}}{\overset{\text{coordination}}{\rightleftharpoons}} H{-}\overset{\oplus}{N}H_3$$

d) Example 3:

$$H_3N{:} \;\; BH_3 \underset{\text{heterolysis}}{\overset{\text{coordination}}{\rightleftharpoons}} H_3\underset{\oplus}{N}{-}\underset{\ominus}{B}H_3$$

Note: In BH_3, B has only six valence electrons and an empty p orbital. It needs two electrons to fill its octet (valence shell). A lone pair on an atom, such as from N in example 3 or from fluoride in example 4, can accomplish filling the octet.

e) Example 4:

$$F^{\ominus} \;\; BH_3 \underset{\text{heterolysis}}{\overset{\text{coordination}}{\rightleftharpoons}} F{-}\underset{\ominus}{B}H_3$$

f) Example 5:

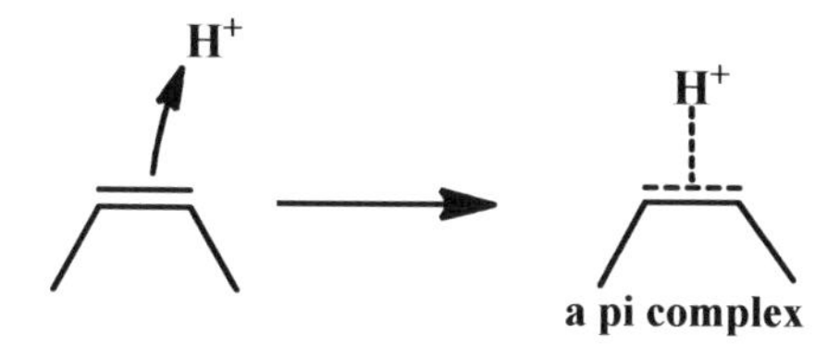

The pi complex can also be drawn like this:

2. Carbocation Rearrangement

This rearrangement will occur when the cation resulting from rearrangement is a more stable carbocation. The greater stability may be, for example, because the rearranged form is a more substituted carbocation (examples 1-5) or because there is less strain in the rearranged carbocation (example 6).

a) General reaction:

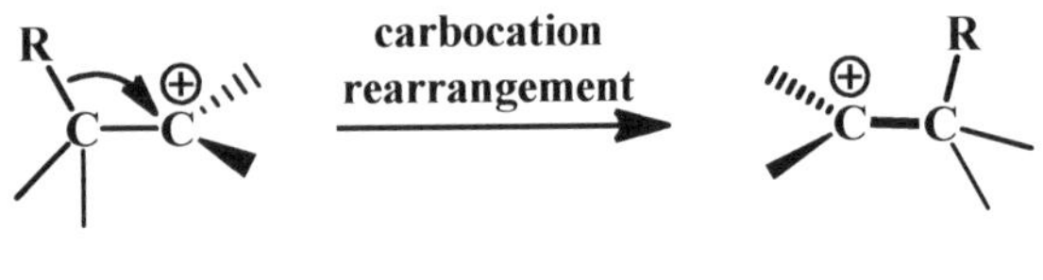

R = H or an alkyl group

b) Example 1: Primary rearranges to secondary carbocation via 1,2-hydride shift

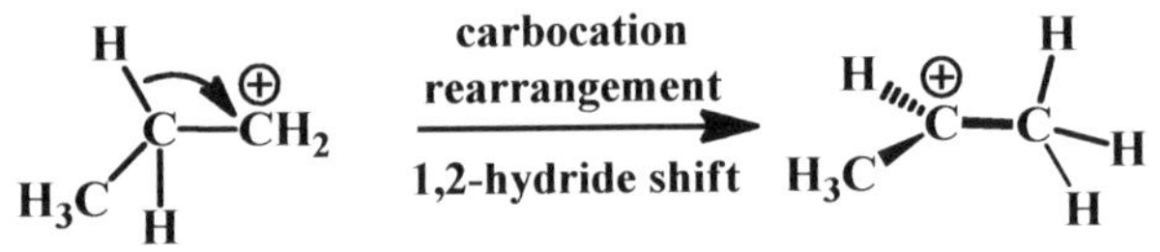

c) Example 2: Primary rearranges to tertiary carbocation via 1,2-hydride shift

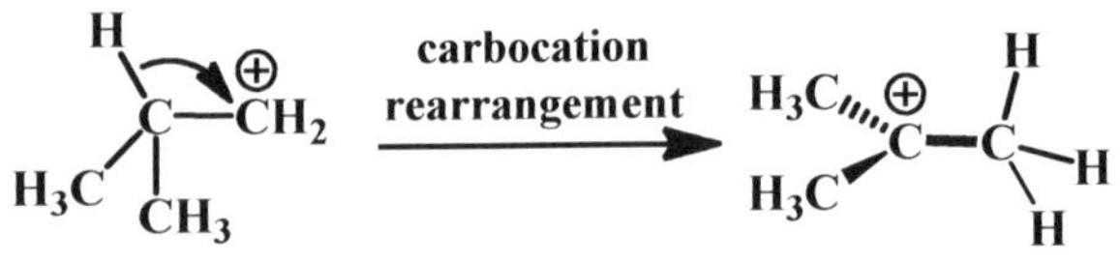

d) Example 3: Secondary rearranges to tertiary carbocation via 1,2-hydride shift

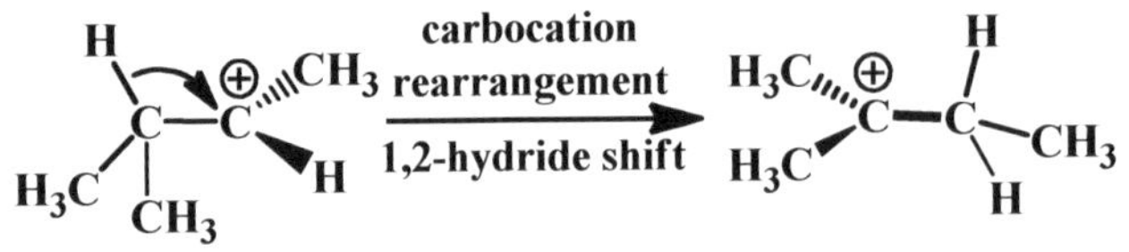

e) Example 4: Primary rearranges to tertiary carbocation via 1,2-methyl shift

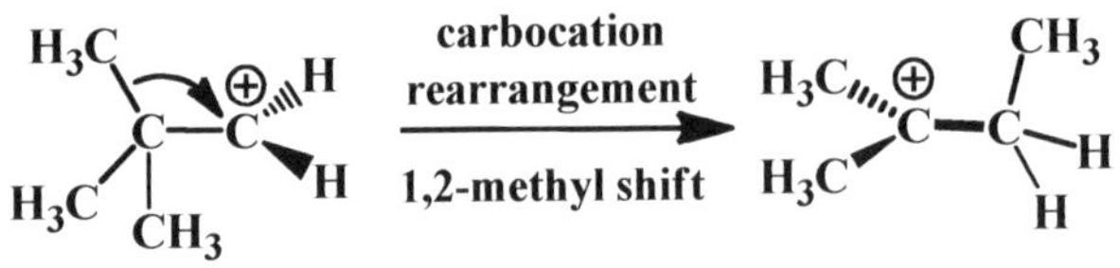

f) Example 5: Secondary rearranges to tertiary carbocation via 1,2-methyl shift

carbocation rearrangement

1,2-methyl shift

g) Example 6: Four-membered ring to five-membered ring to relieve ring strain

carbocation rearrangement

1,2-alkyl shift

In this example, the atoms are numbered to help you see where the carbon atoms in the original structure end up in the rearranged product.

3. Electrophilic Addition / Elimination

In electrophilic addition, pi bond electrons are donated to an electrophile, typically leaving a positive charge on the atom of the pi bond that does not attach to the electrophile. Generally, the more stable cation is formed preferentially, helping one to choose which of the pi-bonded elements will bond to the electrophile and which will have the positive charge on it. The pi bond does not necessarily have to be a C-C pi bond, but it will be for most if not all of the examples you will see in an introductory class.

a) General reaction:

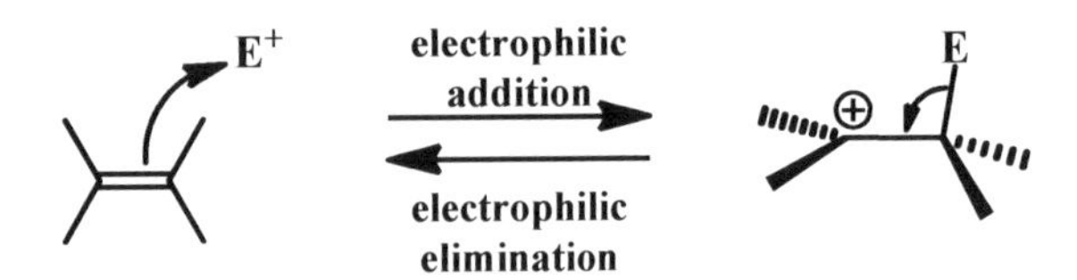

b) Example 1: using a pi bond in an alkene

H^+ — electrophilic addition ⇌ electrophilic elimination

c) Example 2: using a pi bond in a benzene ring

I^+ — electrophilic addition → electrophilic elimination →

This example shows how an electrophilic *addition* of I^+ is followed by electrophilic *elimination* of H^+. The net result of **both** steps is a *substitution* of I for H.

d) Example 3: using a pi bond in an alkyne

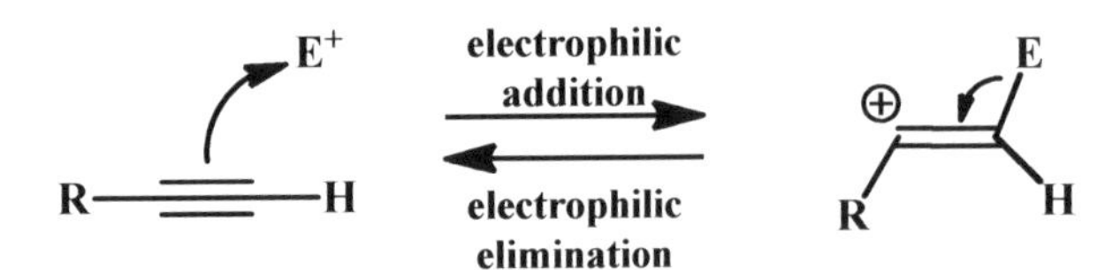

4. The E2 Reaction

In this reaction, a base removes a proton from an sp^2 or sp^3 hybridized atom C1, pushing the electrons from the C1-H bond in between atoms C1 and C2 to form a C1-C2 pi bond. As this happens, one group must leave C2 so that C2 will not have too many electrons (greater than an octet!). All of these electron movements happen at one time, so this is one step; it cannot be broken into separate steps. This reaction is such an important part of organic chemistry that it is covered separately in Section IV.

a) General reaction:

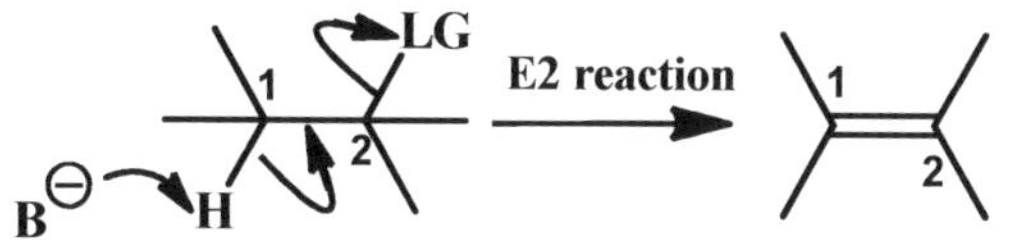

LG E2 reaction + **BH** + **LG**$^{\ominus}$

b) Example 1:

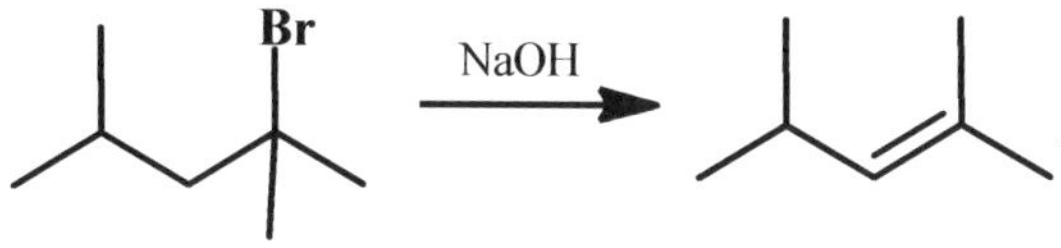

Br NaOH + H_2O + NaBr

c) Example 2:

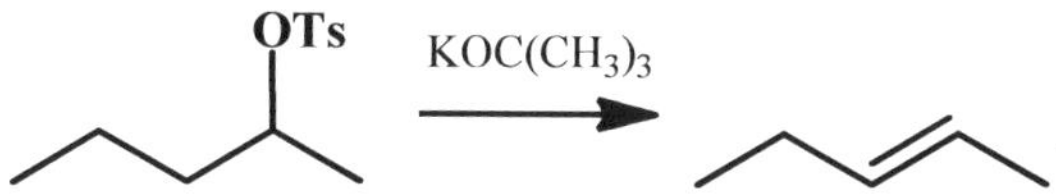

OTs $KOC(CH_3)_3$ + $HOC(CH_3)_3$ + KOTs

trans isomer

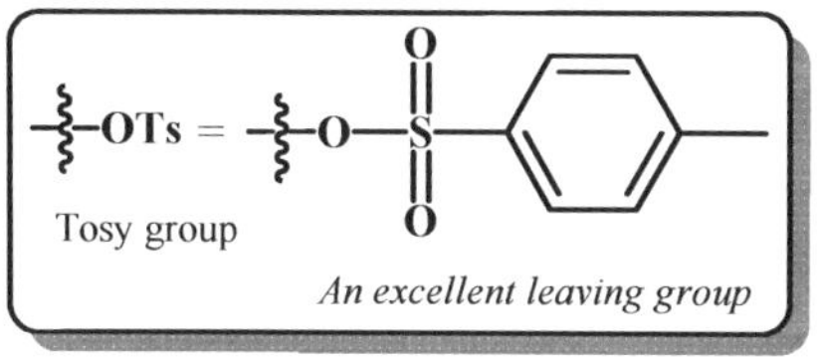

-OTs = -O-S(=O)(=O)-

Tosy group

An excellent leaving group

d) Example 3:

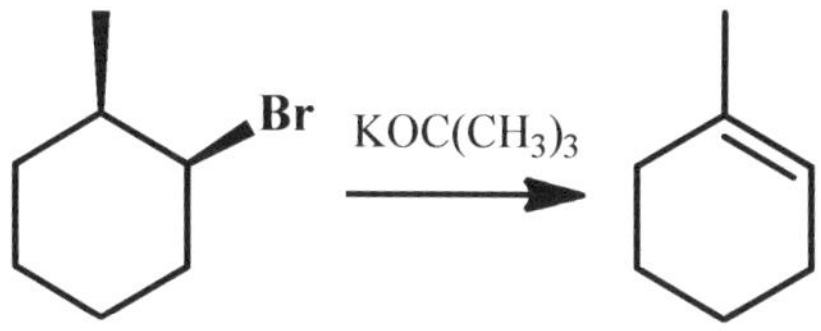

Br $KOC(CH_3)_3$ + $HOC(CH_3)_3$ + KBr

5. The S_N2 Reaction

In this reaction, nucleophile attacks an sp^3 hybridized atom at the partial positive (δ^+) end of a bond. The atom that is attacked by the nucleophile must lose a group (the leaving group) so that the attacked atom doesn't end up with greater than an octet of electrons. All of these electron movements happen at one time, so this is one step; it cannot be broken into separate steps. **This reaction ONLY works when nucleophilic attack is on an sp^3 hybridized atom!**

This reaction is such an important part of organic chemistry that it is covered separately in Section IV.

a) General reaction:

Nu⊖ + CH_3–LG —S_N2 reaction→ Nu–CH_3 + LG⊖

b) Example 1:

KBr → Br + KI

c) Example 2:

Br — $NaOC(O)CH_3$ → + NaBr

d) Example 3:

OTs — $NaSCH_3$ → SCH_3 + NaOTs

–OTs = –O–SO_2–C_6H_4–CH_3

Tosy group

An excellent leaving group

6. Nucleophilic Addition / Elimination

<u>In nucleophilic addition:</u> a nucleophile donates electrons to the δ^+ end of a polar pi bond (in the general reaction shown below, that is atom Y). The electrons that were in the pi bond are then on atom X as a lone pair.

<u>In nucleophilic elimination:</u> electrons from atom X move to an atom Y to make a pi bond between atoms X and Y. The best leaving group on Y has to leave as this pi bond forms so that Y will not have too many bonds (more than an octet of electrons).

a) General reaction:

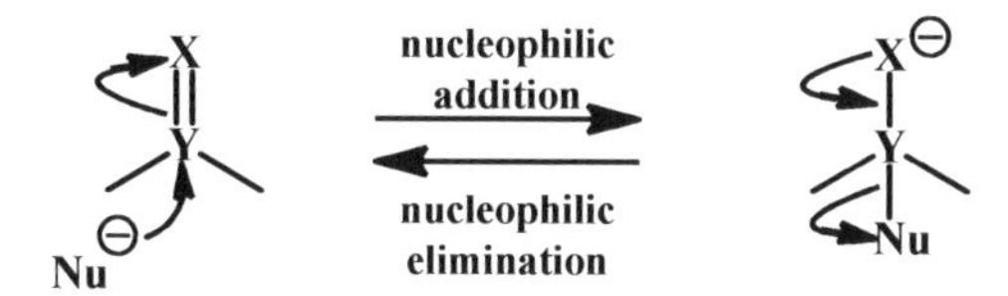

b) Nucleophilic <u>Addition</u> Example 1:

c) Nucleophilic <u>Addition</u> Example 2:

d) Nucleophilic <u>Elimination</u> Example 1:

e) Nucleophilic <u>Elimination</u> Example 2:

II. Reactions of Alkenes and Conjugated Dienes

1. Hydrohalogenation of Alkenes with HX (X = Cl, Br, I)

a) General Net Reaction:

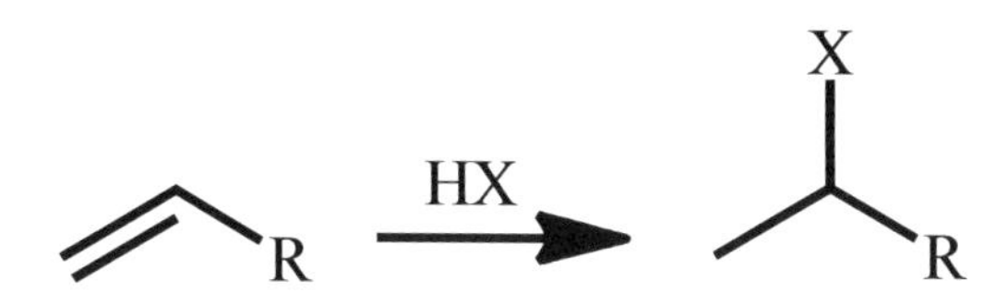

b) Notable Points

- Markovnikov addition (X on the more substituted carbon)
- Essentially 1:1 mixture of *syn*- and *anti*- addition products
- A carbocation intermediate is involved and can rearrange
- Stronger acids react faster (HI > HBr > HCl)

c) Section Covered in Text:

d) Date(s) Discussed in Class

e) Arrow-Pushing Mechanism

i) No Carbocation Rearrangement Needed

H^+ A X^- B X H

ii) With Carbocation Rearrangement

H H^+ C H D X^- E X

iii) Stereochemistry note: addition of a nucleophile to a cation

At this point, it is worth pointing out two additional aspects of this reaction. First, when the alkene gives electrons from its pi bond to the proton, the cation will preferentially form on the more substituted side of the alkene. In the case below, this will be at carbon A:

H^+ C_2H_5 H carbon Ⓐ H H carbon Ⓑ C_2H_5 + CH_3 H carbon Ⓐ carbon Ⓑ

Once the cation is formed, carbon with the positive charge (A in this case) will be sp^2 hybridized, and the geometry about the carbon is trigonal planar. Because the geometry is trigonal planar, the top and bottom portion of the empty p orbital are identical, meaning that a nucleophile has an equal probability of adding to either side:

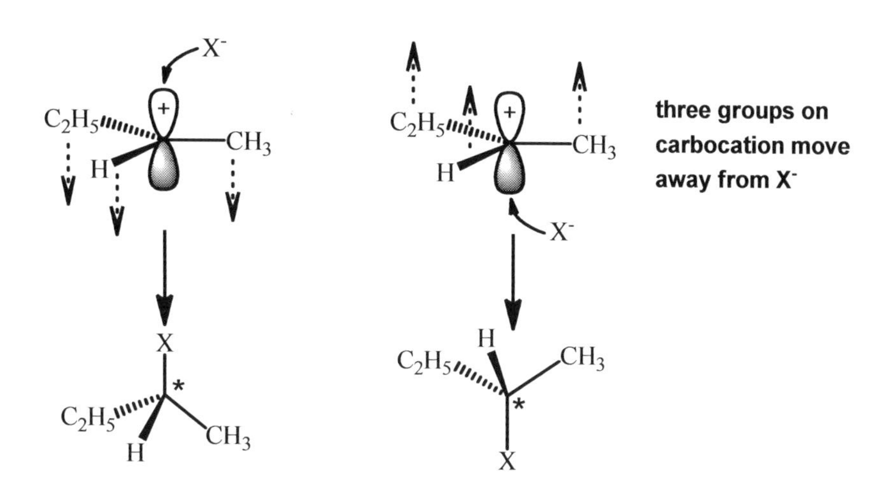

Note that this means that if the addition of the nucleophile to the carbocation produces a stereocenter, the product will be isolated as a racemic mixture (i.e., an equal mixture of the two enantiomers shown in the reaction above).

f) Self Test

i) Provide the name for each mechanistic step shown on the previous page (choices for mechanistic steps are given on page 8).

Step A = Electrophilic Addition

Step B = coordination

Step C = Electrophilic Addition

Step D = carbocation Rearrangement

Step E = coordination

ii) Is net addition of HX Markovnikov or anti-Markovnikov?

Markovnikov

iii) Is addition of HX predominantly *syn-*, *anti-*, or an essentially equal mixture of both? It is an equal mixture

iv) What is the most important intermediate in this reaction (or is the reaction concerted)? carbocation intermediate and rearrangement

v) Can the structure of the intermediate rearrange during the reaction?

Yes!

vi) Draw the major product(s) of these reactions:

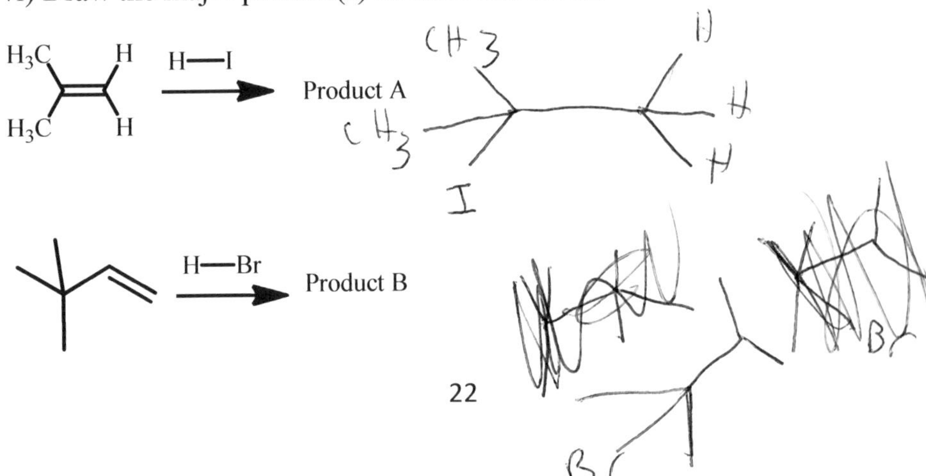

g) Answers to Self Test

i) Mechanistic Steps

Step A = Electrophilic addition

Step B = Coordination

Step C = Electrophilic addition

Step D = Carbocation rearrangement

Step E = Coordination

ii) Markovnikov

iii) An essentially equal mixture of both

iv) A carbocation

v) Yes

vi) Major product(s):

Product A

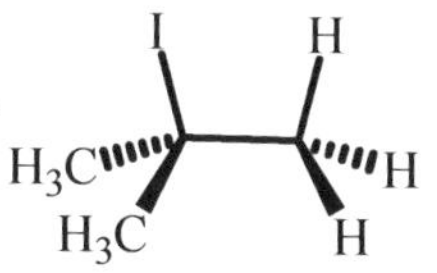

Product B

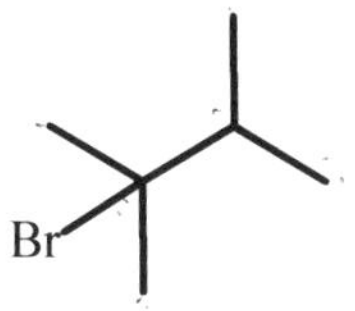

2. Addition of water (hydration) or alcohols to alkenes

a) General Net Reaction:

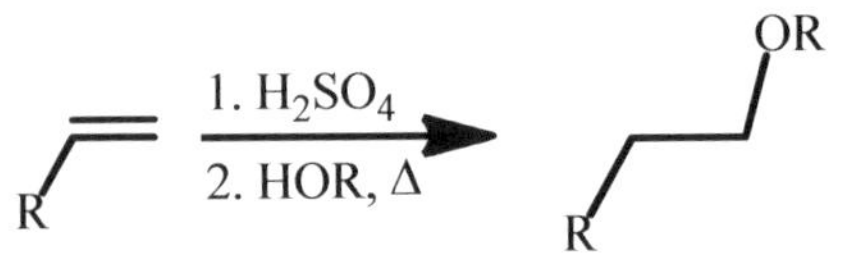

$$R\text{-}CH{=}CH_2 \xrightarrow[\text{2. HOR, }\Delta]{\text{1. }H_2SO_4} R\text{-}CH_2\text{-}CH_2\text{-}OR$$

(HOR; R= H for water, alkyl group for simple alcohols)

b) Notable Points

- Markovnikov addition (-OR on the more substituted carbon)
- Essentially 1:1 mixture of *syn*- and *anti*- addition products
- A carbocation intermediate is involved and can rearrange

c) Section Covered in Text:

d) Date(s) Discussed in Class

e) Arrow-Pushing Mechanism

i) General Mechanism

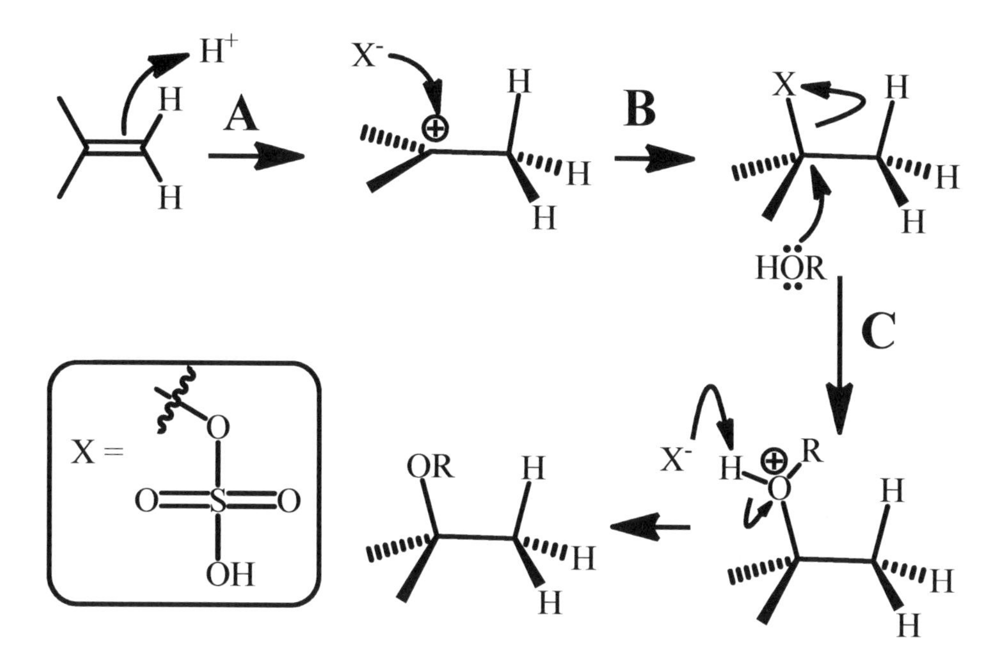

Note that, prior to coordination of X, the intermediate carbocation formed after step A will rearrange if rearrangement leads to a more stable carbocation.

iii) Stereochemistry note: addition of a nucleophile to a cation

At this point, it is worth pointing out two additional aspects of this reaction. First, when the alkene gives electrons from its pi bond to the proton, the cation will preferentially form on the more substituted side of the alkene. In the case below, this will be at carbon A:

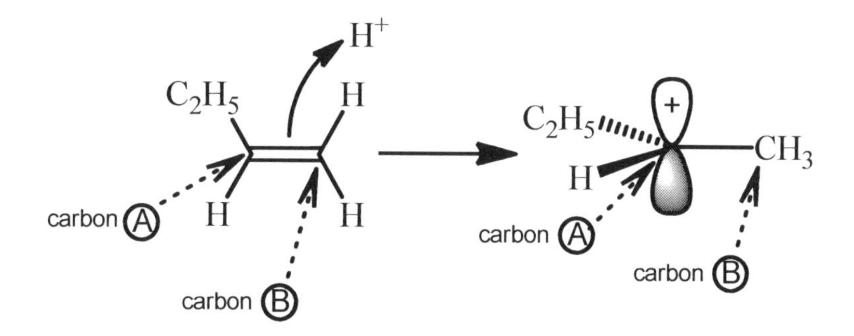

Once the cation is formed, carbon with the positive charge (A in this case) will be sp^2 hybridized, and the geometry about the carbon is trigonal planar. Because the geometry is trigonal planar, the top and bottom portion of the empty p orbital are identical, meaning that a nucleophile has an equal probability of adding to either side:

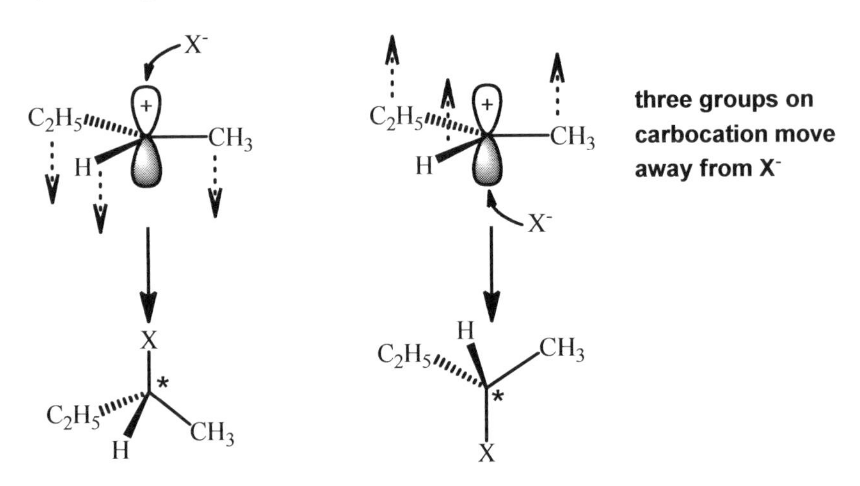

Note that this means that if the addition of the nucleophile to the carbocation produces a stereocenter, the product will be isolated as a racemic mixture (i.e., an equal mixture of the two enantiomers shown in the reaction above).

f) Self Test

i) Provide the name for each mechanistic step shown on the previous page (choices for mechanistic steps are given on page 8).

Step A = Electrophilic Addition

Step B = Coordination

Step C = S_N2

ii) Is net addition of ROH Markovnikov or anti-Markovnikov?

Markovnikov

iii) Is addition of ROH predominantly *syn-*, *anti-*, or an essentially equal mixture of both?

Equal

iv) What is the most important intermediate in this reaction (or is the reaction concerted)?

carbocation

v) Can the structure of the intermediate rearrange during the reaction?

yes

vi) Draw the major product(s) of these reactions:

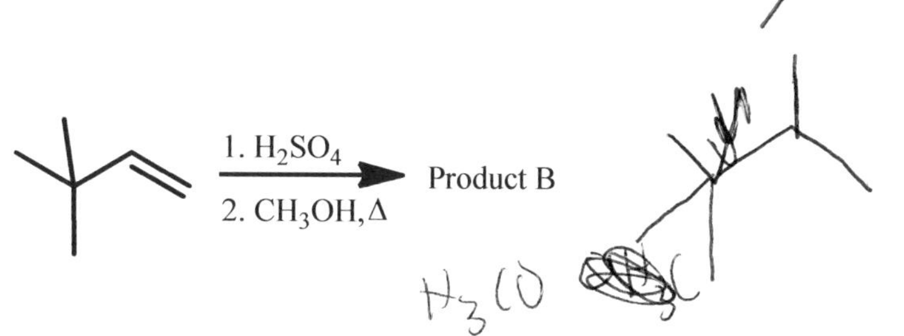

g) Answers to Self Test

i) Provide the name for each mechanistic step shown on the previous page (choices for mechanistic steps are given on page 8).

Step A = Electrophilic addition

Step B = Coordination

Step C = S_N2

ii) Markovnikov

iii) An essentially equal mixture of both

iv) A carbocation

v) Yes

vi) Major products:

Product A

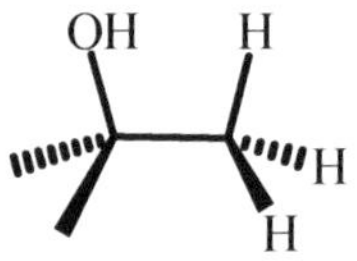

OH H H H

Product B H_3CO

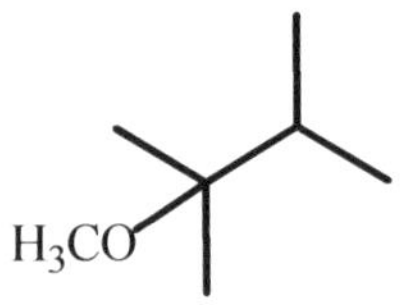

3. Halogenation of alkenes

a) General Net Reaction:

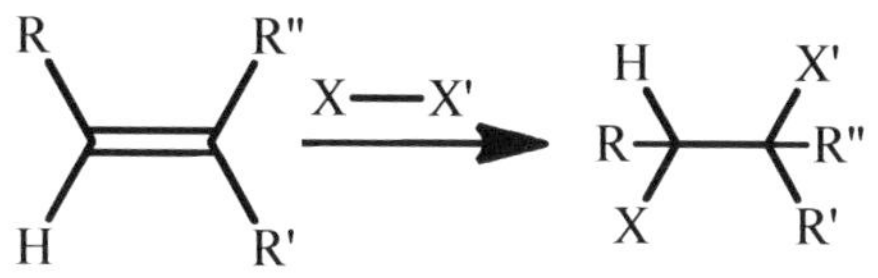

($X\text{-}X' = Cl_2$, Br_2 or Br-Cl)

b) Notable Points

- Markovnikov addition if the two halogens are different. This means that the more electronegative halogen ends up on the more substituted carbon if the two halogens are different (like in Br-Cl)
- Produces the *anti*- addition product
- A halonium intermediate is involved
  - Bromonium when Br_2 or Br-Cl adds
  - Chloronium when Cl_2 adds
- Halonium ions cannot rearrange

c) Section Covered in Text:

d) Date(s) Discussed in Class

e) Arrow-Pushing Mechanism (clarification for step **A** is provided on the next page)

i) For Cl_2 or Br_2 addition:

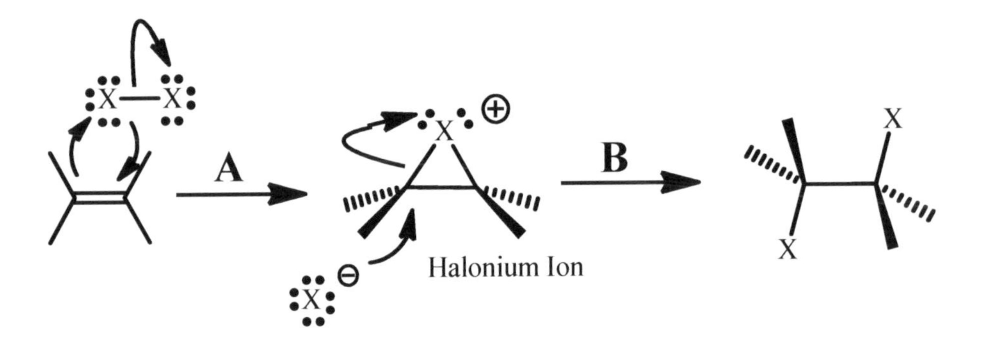

ii) For Cl-Br addition

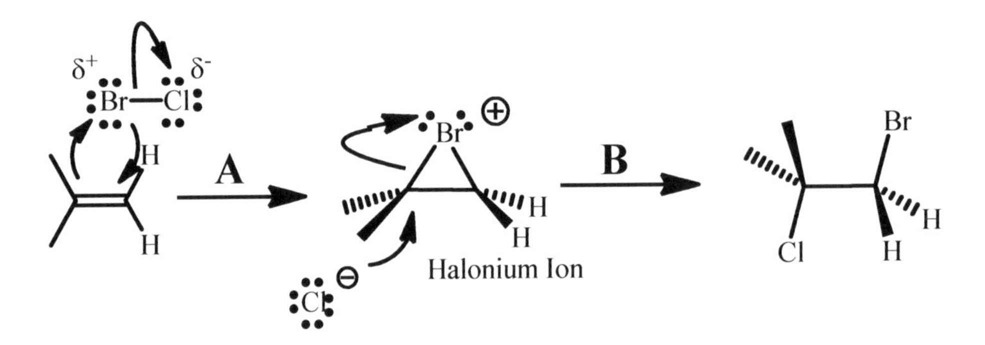

iii) Clarification on step **A** of the mechanisms

Step A is a special mechanistic step that is a combination of **two** of the six basic steps that we have been discussing for mechanisms happening *at the same time*. In the picture below, you can see how we can think about step A as a combination of electrophilic addition and coordination. These two steps do **not** each happen separately; they happen at the same time, so **you never get a carbocation intermediate**. Only a halonium intermediate forms, and it cannot rearrange.

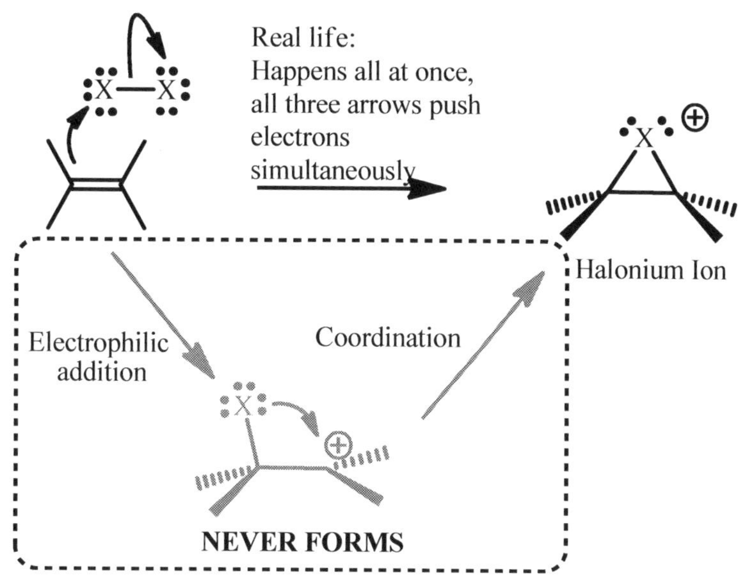

This box shows you how to visualize the two-steps-at-once process; but remember that these steps happen all at once, not individually.

f) Self Test

i) Provide the name for each mechanistic step shown on the previous page (choices for mechanistic steps are given on page 8).

Step A is not a combination of: mechanistic steps

and 6 basic steps, happening in a concerted fashion.

Step B = S_N2

ii) Is net addition of Cl-Br Markovnikov or anti-Markovnikov? Markovnikov

iii) Is net addition of X-X' predominantly *syn*-, *anti*-, or an essentially equal mixture of both? ANTI!

iii) What is the most important intermediate in this reaction (or is the reaction concerted)? Bromonium ion (or Halonium)

iv) Can the structure of the intermediate rearrange during the reaction? NO!

v) Draw the major product(s). Be sure to indicate the stereo and regiochemistry where necessary.

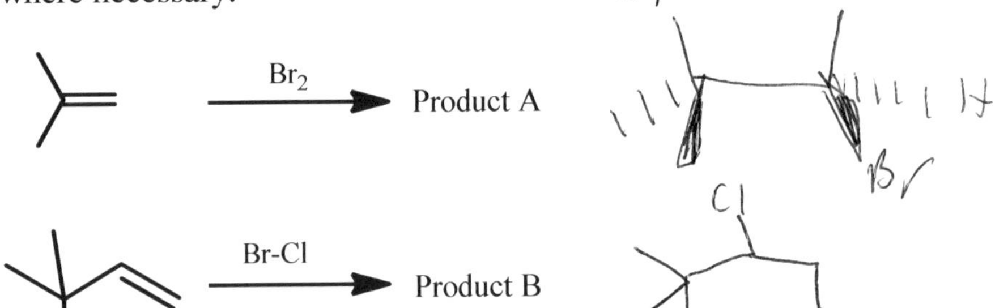

Cl_2 → Product C

Cl

Cl

g) Answers to Self Test

i) Mechanistic Steps

Step A is not one of the six basic steps. It is a combination of electrophilic addition and intramolecular coordination, happening in a concerted fashion.

Step B = S_N2

i) Net addition is Markovnikov.

ii) Net addition is *anti-* .

iii) A halonium (bromonium or chloronium) intermediate is involved.

iv) The intermediate **cannot** undergo rearrangement.

v) Major product(s):

Product A

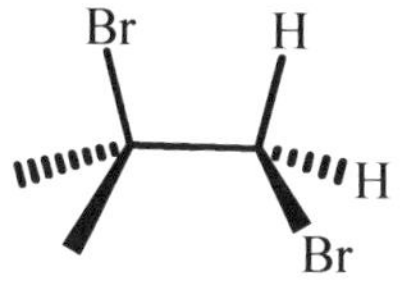

Product B

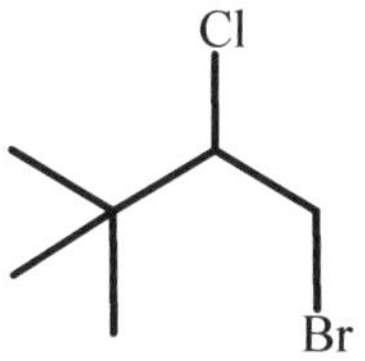

Product C

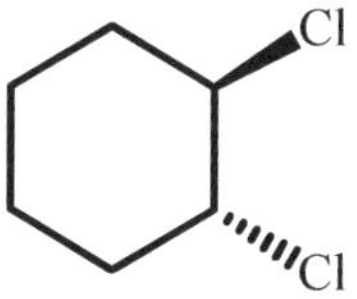

4. Halohydrin formation from alkenes

a) General Net Reaction:

R''' R''
H R'
X—X
HOR
H OR
R''' R''
X R'

(X-X is generally Cl_2 or Br_2. In HOR, R= H for water, alkyl group for simple alcohols)

b) Notable Points

- Markovnikov addition; The –OR ends up on the more substituted carbon.
- Produces the *anti-* addition product
- A halonium intermediate is involved
- Halonium ions cannot rearrange
- In practical settings, the water or alcohol is used in excess so that addition of the –OR group is favored over halogenation

c) Section Covered in Text:

d) Date(s) Discussed in Class

e) Arrow-Pushing Mechanism

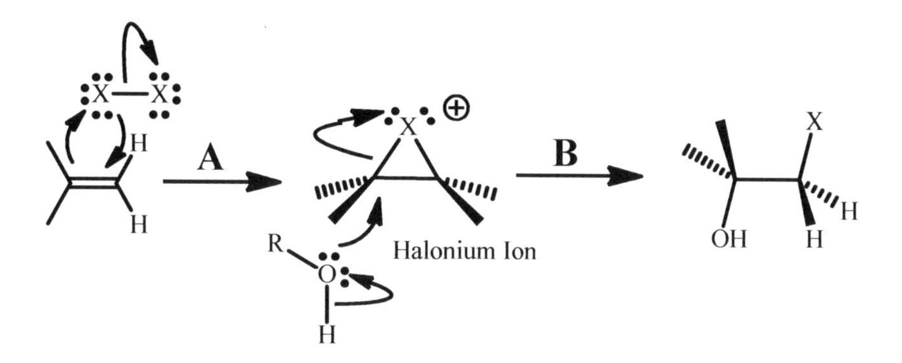

For clarification on step **A** of the mechanisms, see reaction II.3.e.iii. (part e of the previous reaction in this text).

f) Self Test

i) Provide the name for each mechanistic step shown on the previous page (choices for mechanistic steps are given on page 8).

Step A is a combination of: Electrophilic Addition

and Intramolecular coordination, happening in a concerted fashion.

Step B = SN2

ii) Is net addition of -OR and -X groups Markovnikov or anti-Markovnikov?

iii) Is net addition of -OR and -X groups predominantly *syn*-, *anti*-, or an essentially equal mixture of both? Anti

iii) What is the most important intermediate in this reaction (or is the reaction concerted)? Halonium

iv) Can the structure of the intermediate rearrange during the reaction? NO!

v) Draw the major product(s). Be sure to indicate the stereo and regiochemistry where necessary.

$\xrightarrow[H_2O]{Br_2}$ Product A

OH H Br

$\xrightarrow[HOCH_2CH_2CH_3]{Br_2}$ Product B

$O(CH_2)_2CH_3$ Br

$\xrightarrow[HOCH_3]{Cl_2}$ Product C

Cl OCH_3 +

g) Answers to Self Test

i) Mechanistic Steps

Step A is not one of the six basic steps. It is a combination of electrophilic addition and intramolecular coordination, happening in a concerted fashion.

Step B = S_N2

i) Net addition is Markovnikov.

ii) Net addition is *anti-* .

iii) A halonium (bromonium or chloronium) intermediate is involved.

iv) The intermediate **cannot** undergo rearrangement.

v) Major product(s):

Product A

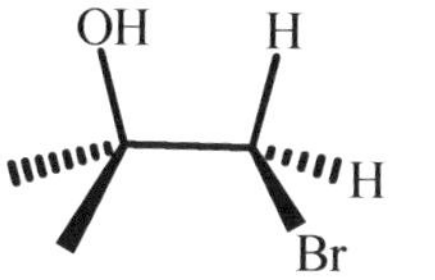

Product B

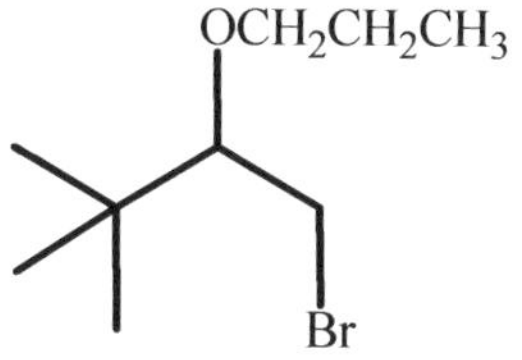

Product C

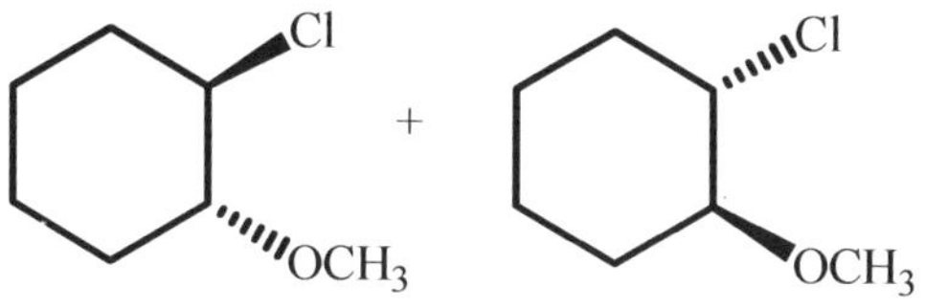

5. Oxymercuration / Reduction of alkenes: Two steps to an alcohol

a) General Net Reaction:

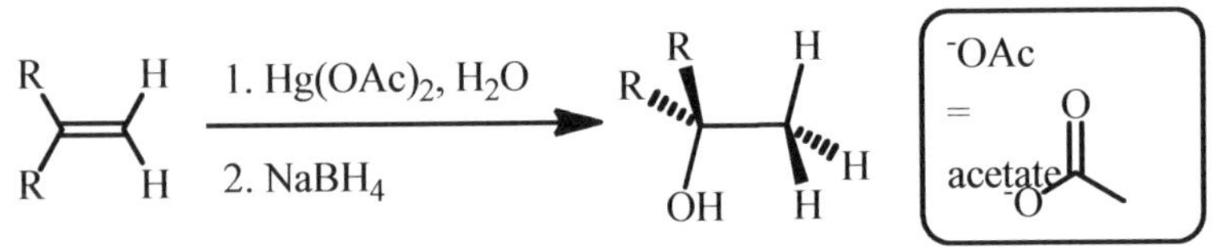

1. $Hg(OAc)_2$, H_2O
2. $NaBH_4$

^{-}OAc = acetate

b) Notable Points

- Net Markovnikov addition (-OH on the more substituted carbon)
- Essentially 1:1 mixture of *syn-* and *anti-* addition products
- No carbocation intermediate is involved
- None of the intermediates involved can undergo a carbocation rearrangement
- Often the best way to make a Markovnikov alcohol when you want to avoid rearrangement or use of strong acid

c) Section Covered in Text:

d) Date(s) Discussed in Class

e) Arrow-Pushing Mechanism

i) Oxymercuration Step

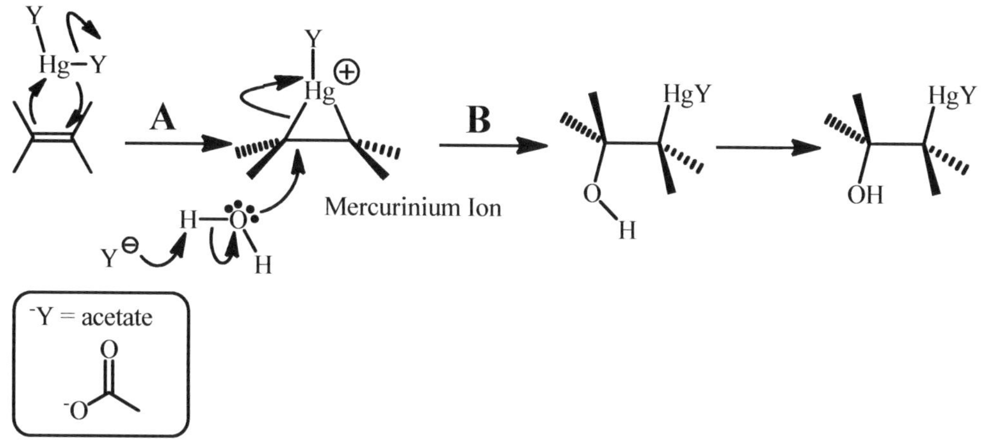

For clarification on step **A** of the mechanisms, see reaction II.3.e.iii.

ii) Reduction Step

The mechanism of the reduction step is somewhat complex, and most instructors do not require you to study it in detail at the introductory level. One reasonable mechanism favored by some researchers is provided here for the interested reader, the involvement of a radical provides you with the mechanistic reason behind why there is no **specificity** for *syn-* or *anti-* addition.

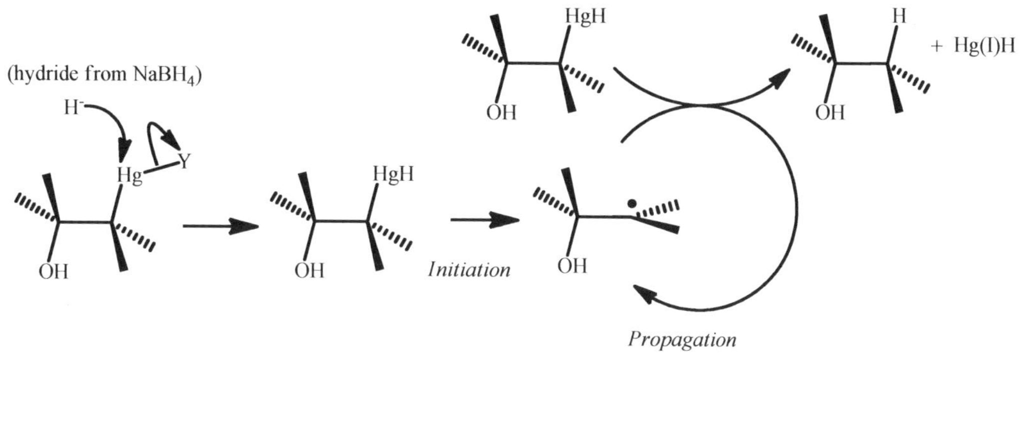

f) Self Test

i) Provide the name for each mechanistic step shown on the previous page (choices for mechanistic steps are given on page 8).

Step A is not a combination of: Electrophilic Addition

and Intramolecular Coordination, happening in a concerted fashion.

Step B = S_N2

ii) Is net addition of -H and -OH Markovnikov or anti-Markovnikov?

Markovnikov

iii) Is net addition of -H and -OH predominantly *syn*-, *anti*-, or an essentially equal mixture of both? Equal Mix

iii) What is the most important intermediate in this reaction (or is the reaction concerted)? Mercurinium

iv) Can the structure of the intermediate rearrange during the reaction?

No!

v) Draw the major product(s):

1. $Hg(OAc)_2$, H_2O / 2. $NaBH_4$ → Product A

OH H H H

1. $Hg(OAc)_2$, H_2O / 2. $NaBH_4$ → Product B

OH

1. $Hg(OAc)_2$, H_2O / 2. $NaBH_4$ → Product C

OH

g) Answers to Self Test

i) Mechanistic Steps

Step A is a combination of electrophilic addition and intramolecular coordination, happening in a concerted fashion.

Step B = S_N2

i) Net addition is Markovnikov.

ii) Net addition is essentially an equal mixture of both

iii) A mercurinium intermediate is involved.

iv) The intermediate **cannot** undergo rearrangement.

v) Major product(s):

Product A

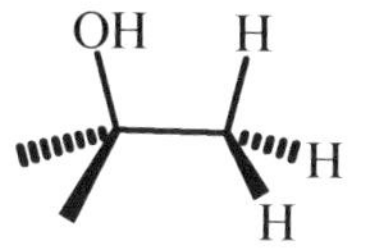

Product B

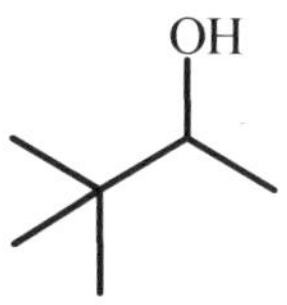

Product C

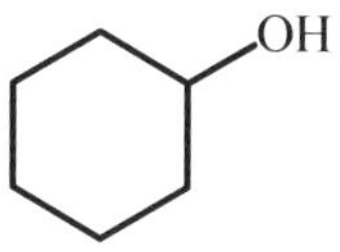

6. Epoxidation of alkenes

a) General Net Reaction:

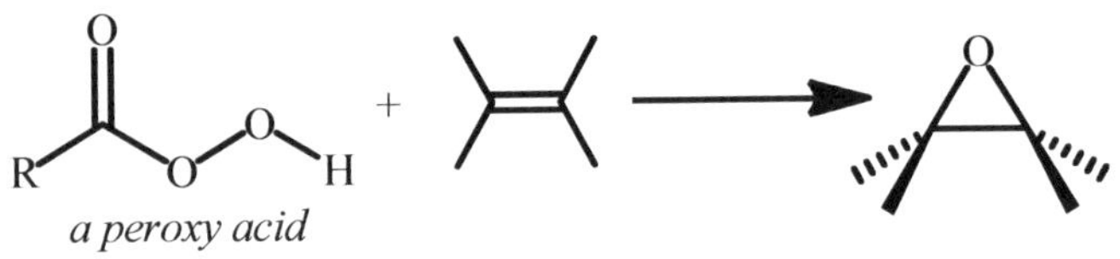

a peroxy acid

b) Notable Points

- The reaction is concerted
- There are no intermediate species
- The carbon skeleton cannot rearrange during the course of reaction
- The face of the alkene to which the O adds is an important consideration for stereochemistry of product in some cases.
- One of the most common peracids used for epoxidations is *m*-chloroperoxybenzoic acid, usually just abbreviated mCPBA:

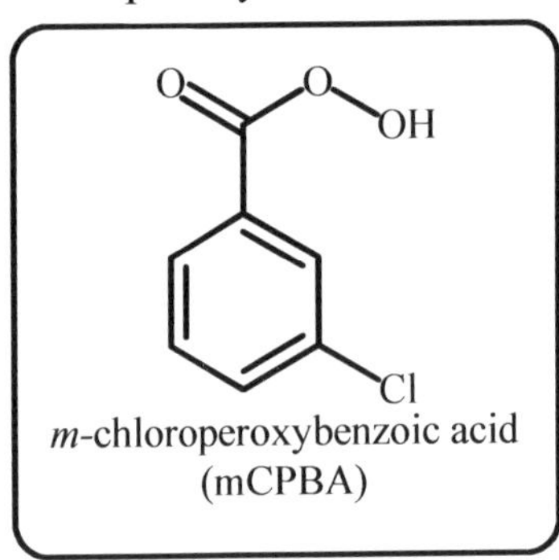

m-chloroperoxybenzoic acid (mCPBA)

c) Section Covered in Text:

d) Date(s) Discussed in Class

e) Arrow-Pushing Mechanism

i) General mechanism

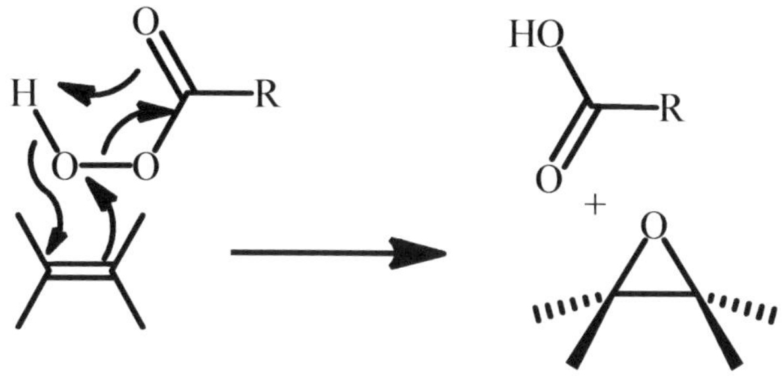

ii) Stereochemistry note

Because epoxidation is a concerted reaction, the alkene substituents **cannot** rearrange with respect to one another during the reaction. An important ramification of this observation is that when chiral centers are produced upon epoxidation, the *face* of the alkene to which the O adds is important. Furthermore, if neither starting material (peracid or alkene) is chiral, there will not be a preference for one of two enantiomers, so both would be produced in a 1:1 ratio:

Addition to 'top' face:

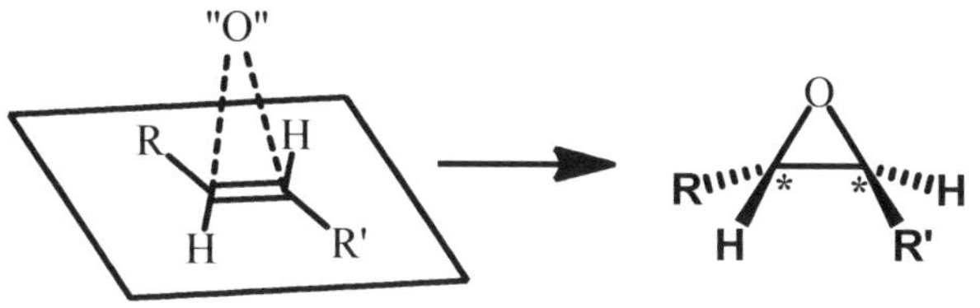

Addition to 'bottom' face:

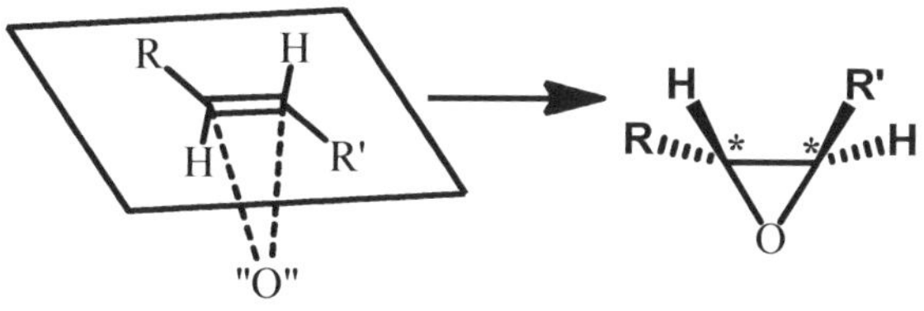

f) Self Test

i) What is the most important intermediate in this reaction (or is the reaction concerted)? concerted

ii) Can the structure of the intermediate rearrange during the reaction?

iii) Draw the major product(s). Be careful to look for potential chiral products and evaluate whether you may need to draw more than one product isomer.

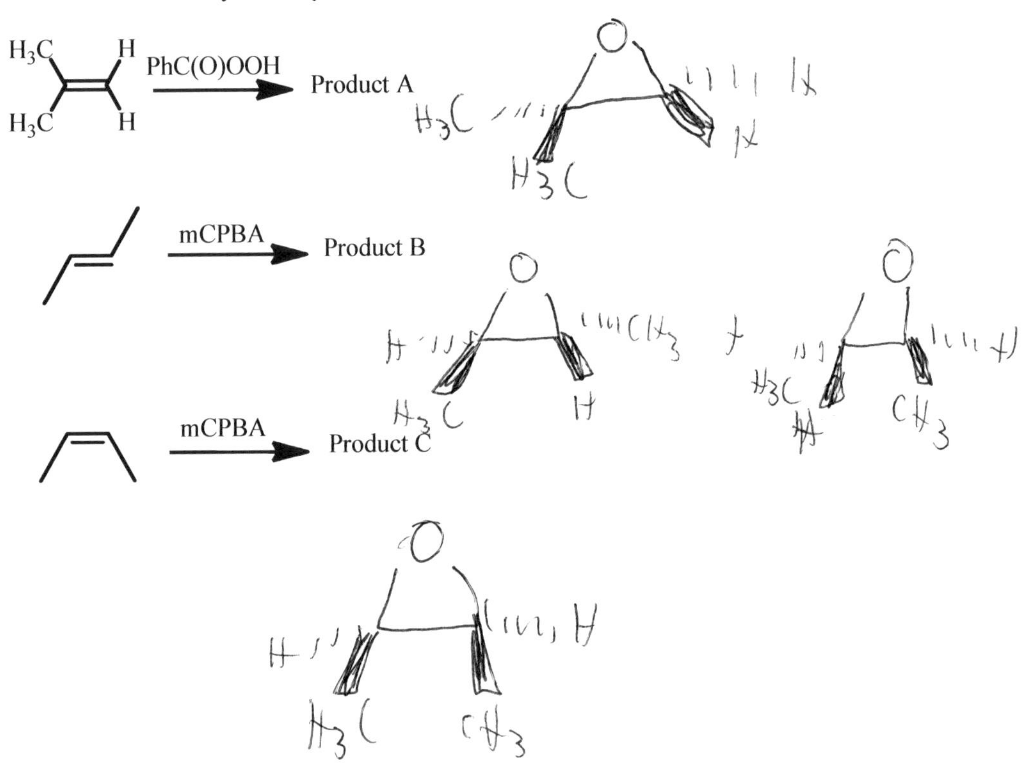

g) Answers to Self Test

i) This is a concerted reaction

ii) no rearrangement can occur

iii) Major product(s):

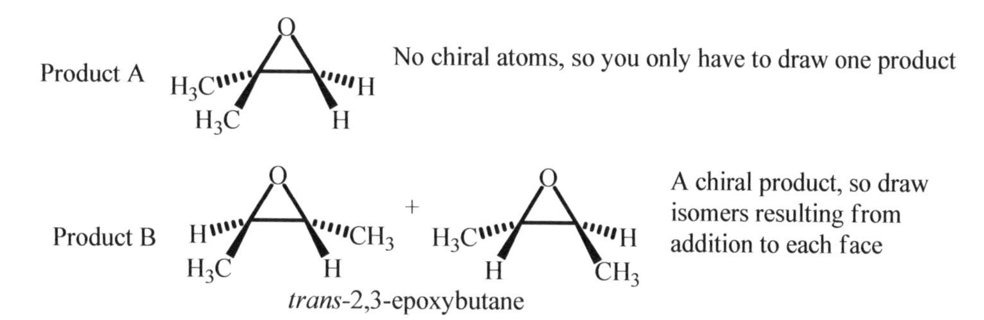

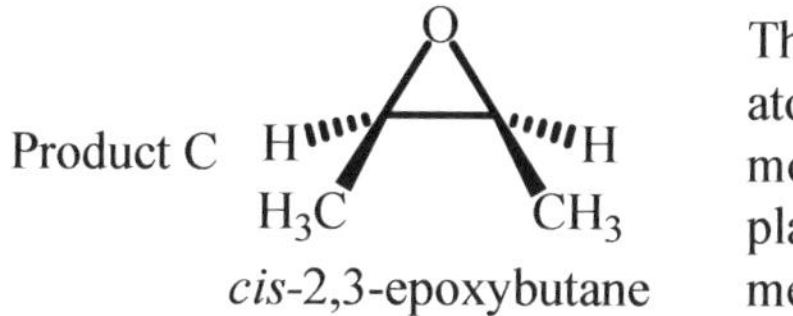

This compound has chiral atoms, but is not a chiral molecule because it has a plane of symmetry; it is a meso compound.

7. Cyclopropanation of alkenes

a) General Net Reaction:

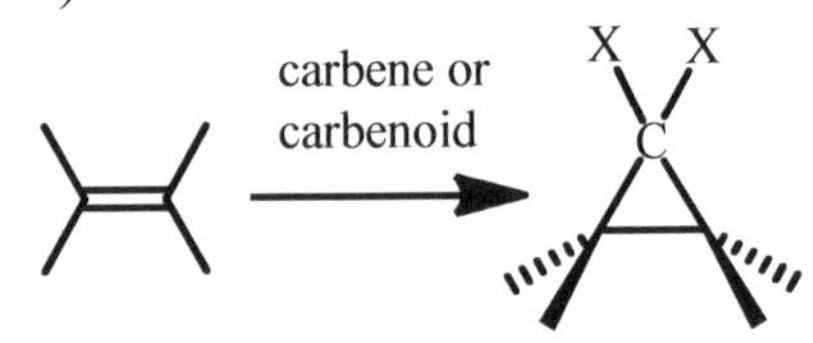

b) A general reaction that employs a carbene:

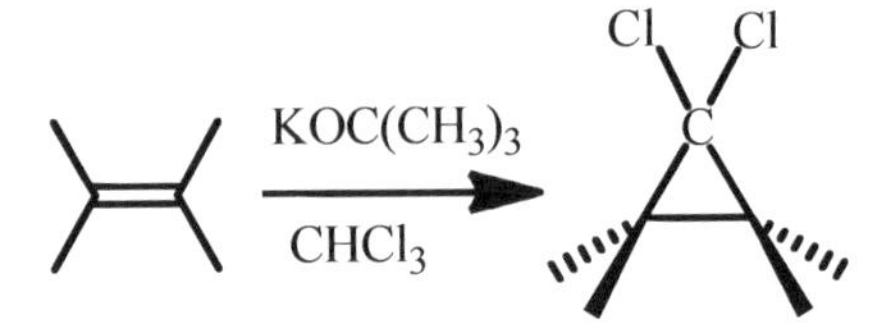

$KOC(CH_3)_3$ plus $CHCl_3$ makes dichlorocarbene:

$KOC(CH_3)_3 + CHCl_3 \longrightarrow :CCl_2 + HOC(CH_3)_3 + KCl$

c) A general reaction that employs a carbenoid (**Simmons-Smith reaction**):

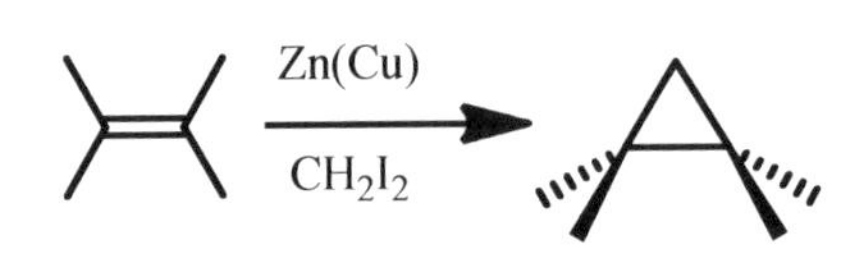

When Zn(Cu) and CH_2I_2 are used, this is called the **Simmons-Smith Reaction**

- Zn(Cu) reacts with CH_2I_2 to make a carbenoid: H H C I ZnI
- A carbeneoid is something that is not a free carbene but can do reactions similar to carbene reactions

d) Notable Points

- The reaction is concerted
- There are no intermediate species
- The carbon skeleton cannot rearrange during the course of reaction
- The face of the alkene to which the CX_2 unit adds is an important consideration for stereochemistry of product in some cases.

e) Section Covered in Text:

f) Date(s) Discussed in Class

g) Arrow-Pushing Mechanism

i) Using a free carbene (the carbene is made before it reacts with the alkene)

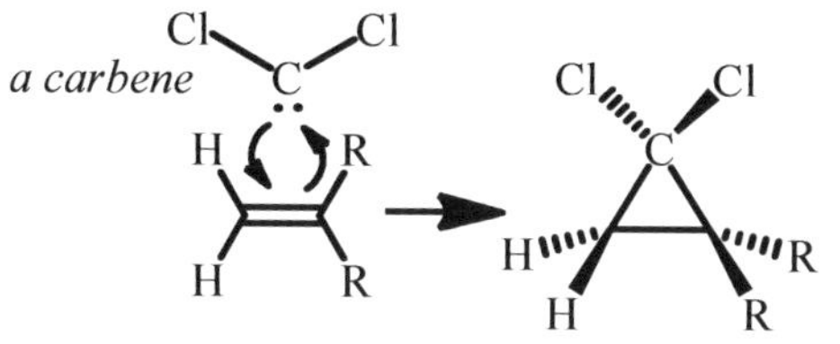

ii) Using a carbenoid (the carbene fragment is produces *while* the species reacts with the alkene)

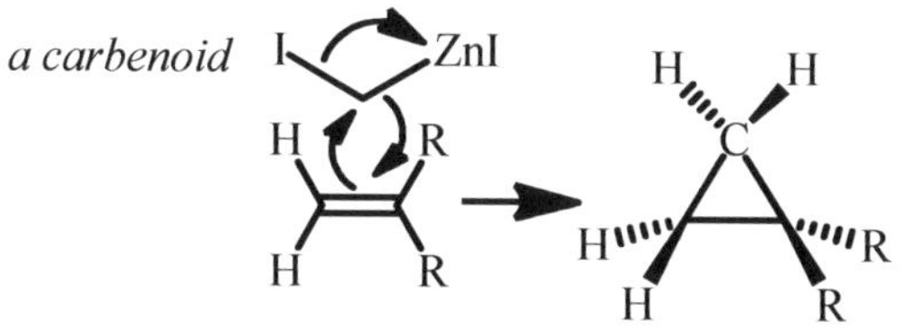

iii) Stereochemistry note

Because epoxidation is a concerted reaction, the alkene substituents **cannot** rearrange with respect to one another during the reaction. An important ramification of this observation is that when chiral centers are produced upon epoxidation, the *face* of the alkene to which the O adds is important. Furthermore, if neither starting material (peracid or alkene) is chiral, there will not be a preference for one of two enantiomers, so both would be produced in a 1:1 ratio:

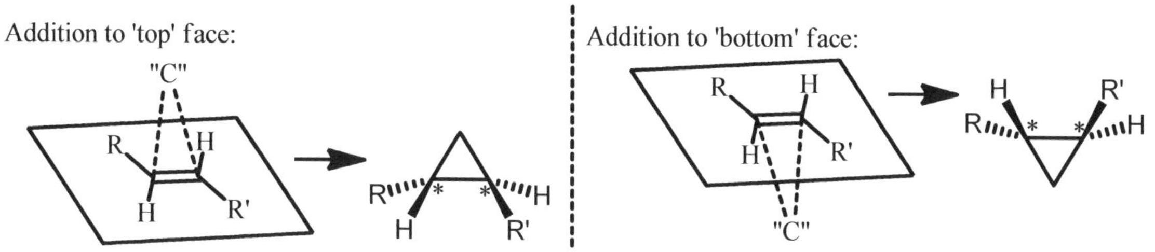

Addition to 'top' face:

Addition to 'bottom' face:

h) Self Test

i) What is the most important intermediate in this reaction (or is the reaction concerted)? Concerted

ii) Can the structure of the intermediate rearrange during the reaction?

Nope.

iii) Draw the major product(s). Be careful to look for potential chiral products and evaluate whether you may need to draw more than one product isomer.

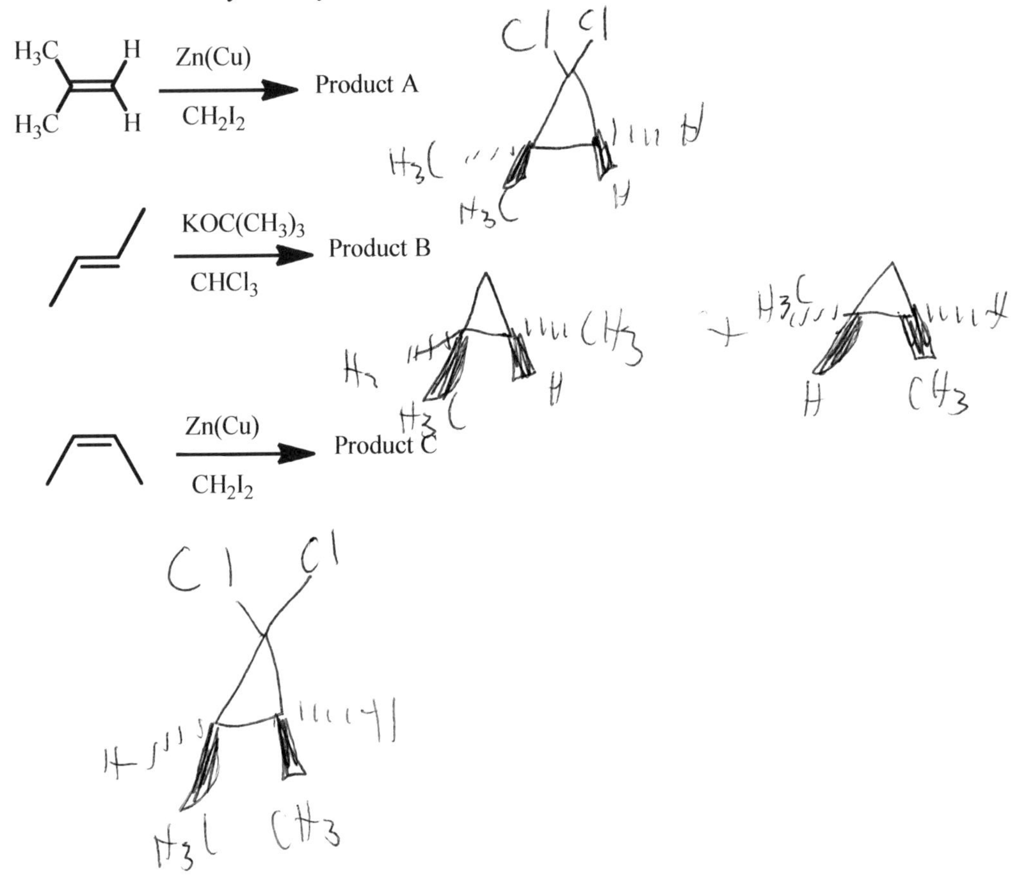

i) Answers to Self Test

i) This is a concerted reaction

ii) no rearrangement can occur

iii) Major product(s):

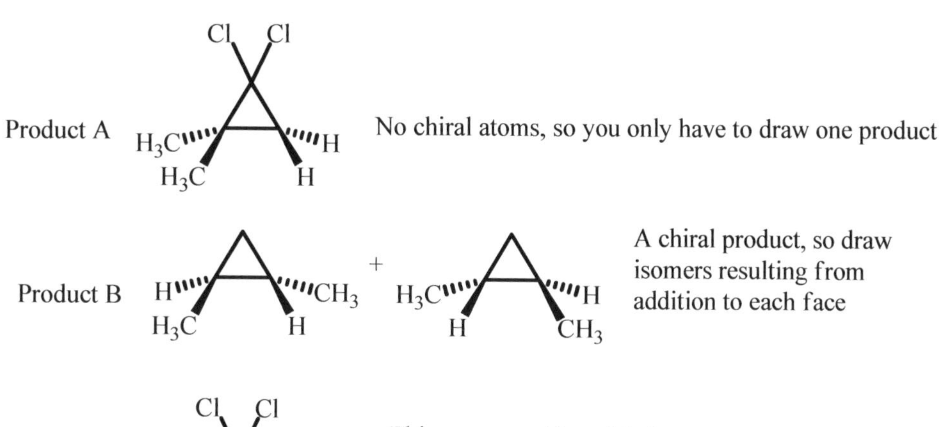

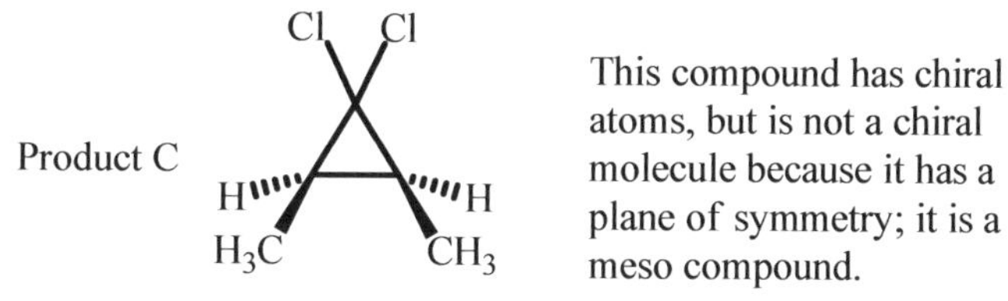

8. Hydroboration / oxidation of alkenes: Two steps to an alcohol

a) General Net Reaction:

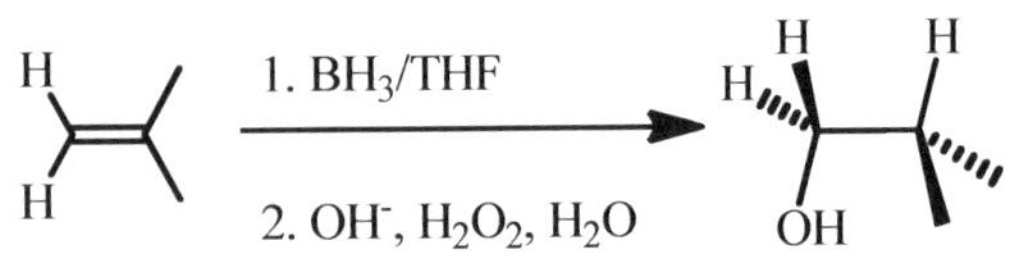

1. BH_3/THF
2. OH^-, H_2O_2, H_2O

(you may see "B_2H_6" or "R_2BH" in place of "BH_3/THF")

b) Notable Points

- Anti-Markovnikov addition; the –OH ends up on the **less** substituted carbon.
- Produces the *syn*- addition product
- No carbocation intermediate is involved
- This is a two-reaction sequence:
  - 1. Hydroboration adds -H and $-BR_2$
  - 2. In oxidation, the $-BR_2$ piece is converted to -OH
- This is the **only** direct route to anti-Markovnikov alcohols taught in most introductory courses.
- This is one of the key anti-Markovnikov addition reactions of alkenes taught in most introductory organic courses (see also Section II, reaction 11.2: Hydrobromination of alkenes in the presence of peroxide,)

c) Section Covered in Text:

d) Date(s) Discussed in Class

e) Arrow-Pushing Mechanism

i) General mechanism

partial positive charge on carbon to which H adds

syn addition

"anti-Markovnikov Product"

ii) Stereochemistry note

Because the hydroboration step is concerted, the alkene substituents **cannot** rearrange with respect to one another during the reaction. The stereochemistry of the product is set in the hydroboration step. An important ramification of this observation is that when chiral centers are produced during hydroboration/oxidation, the *face* of the alkene to which the –H and –BR_2 pieces add is important. Furthermore, if neither starting material (HBR_2 or alkene) is chiral, there will not be a preference for one of two enantiomers, so both would be produced in a 1:1 ratio:

Addition to 'top' face:

Addition to 'bottom' face:

f) Self Test

i) Is addition of –H and –OH a Markovnikov or anti-Markovnikov addition?

Anti-Makov!

ii) Is addition of –H and -OH predominantly *syn-*, *anti-*, or an essentially equal mixture of both?

Syn

iii) Is there an intermediate that can rearrange the carbon skeleton of the starting material during reaction?

No!

iv) What is added to the alkene after the hydroboration step before oxidation?

H and BR_2

v) What transformation does the oxidation step perform?

The BR_2 piece converts to OH

vi) Draw the major product(s):

1. BH_3/THF
2. OH^-, H_2O_2, H_2O → Product A

1. BH_3/THF
2. OH^-, H_2O_2, H_2O → Product B

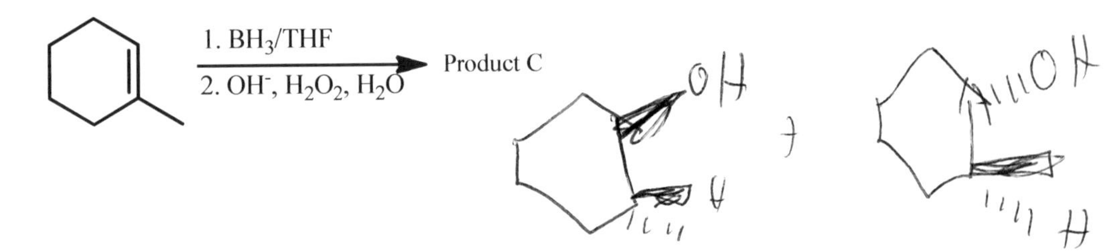

g) Answers to Self Test

i) Anti-Markovnikov addition?

ii) Net addition of –H and -OH is *syn*-.

iii) There are no intermediates that can rearrange the carbon skeleton.

iv) Hydroboration adds -H and $-BR_2$

v) In oxidation, the $-BR_2$ piece is converted to -OH

vi) Major product(s):

Product A

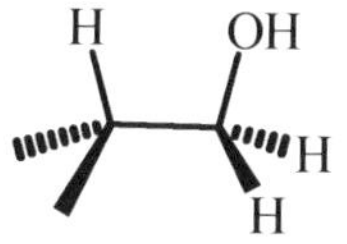

H OH H H

Product B

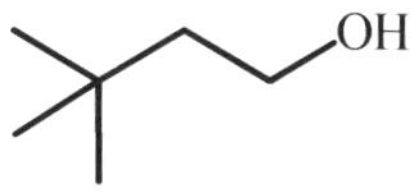

OH

Product C

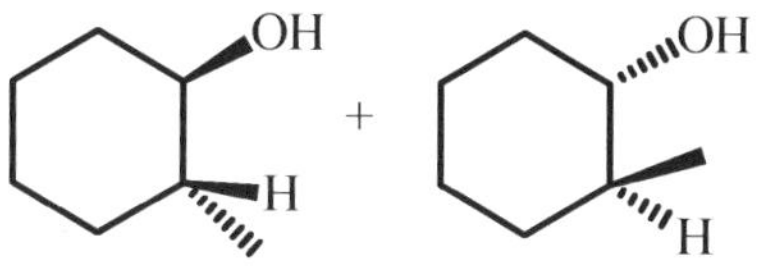

OH H + OH H

9. Oxidation of alkenes to vicinal diols

a) General Net Reaction:

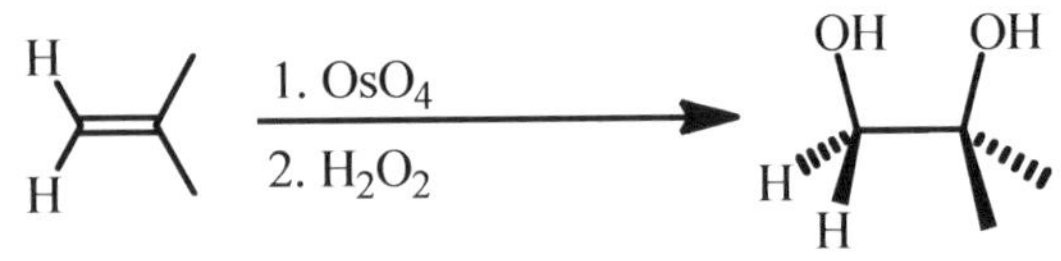

1. OsO_4
2. H_2O_2

b) Notable Points

- Both groups added are the same, so you do not have to differentiate between Markovnikov or anti-Markovnikov addition
- Produces the *syn*- addition product
- No carbocation intermediate is involved
- The carbon skeleton does not rearrange during reaction
- This is a two-reaction sequence:
    - 1. OsO_4 adds in a concerted step
    - 2. Hydrogen peroxide (H_2O_2) converts the osmium piece to the two –OH units.

c) Section Covered in Text:

d) Date(s) Discussed in Class

e) Arrow-Pushing Mechanism

i) Mechanism

H_2O_2

syn addition

(arrow-pushing not provided for this step)

ii) Stereochemistry note

Because the OsO_4 addition step is concerted, the alkene substituents **cannot** rearrange with respect to one another during the reaction, so the stereochemistry of the product is set in the first step shown above. As with several other reactions we have examined so far in this text, when chiral centers are produced during this reaction, the *face* of the alkene to which the OsO_4 fragment adds is important. Furthermore, since OsO_4 is achiral, there will not be a preference for one of two enantiomers, so both would be produced in a 1:1 ratio:

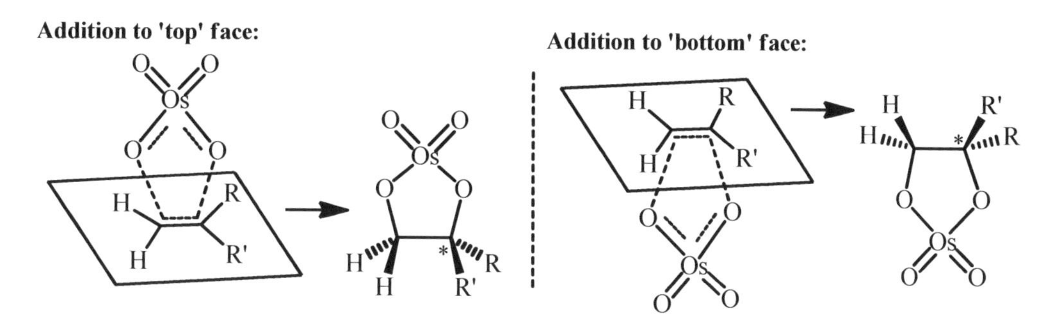

we didnt cover in class. so I will copy the answers

f) Self Test

i) Is addition of the two –OH groups a Markovnikov or anti-Markovnikov addition?

neither

ii) Is addition of the two –OH groups predominantly *syn*-, *anti*-, or an essentially equal mixture of both?

syn

iii) Is there an intermediate that can rearrange the carbon skeleton of the starting material during reaction? NO

iv) What is the structure of the product resulting from reaction of the alkene with OsO_4 before reaction with other reagents?

5-membered ring w/ 2 alkene C, O, Osmium

v) What transformation does the H_2O_2 perform?

Removes osmium-containing portion, leaves an OH on each C

vi) Draw the major product(s):

1. OsO_4 2. H_2O_2 → Product A

1. OsO_4 2. H_2O_2 → Product B

1. OsO_4 2. H_2O_2 → Product C

g) Answers to Self Test

i) This is a trick question! Since both groups are the same, it is not Markovnikov or anti-Markovnikov.

ii) Specific *syn*-addition is observed

iii) There are no intermediates that can rearrange the carbon skeleton

iv) A five-membered ring with the two alkene carbons, two oxygens, and an osmium:

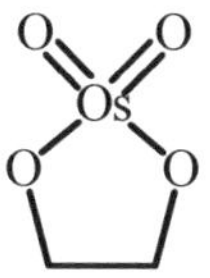

O O
Os
O O

v) It removes the osmium-containing portion and leaves a –OH on each carbon to which the Os-containing piece was attached.

vi) major product(s):

Product A

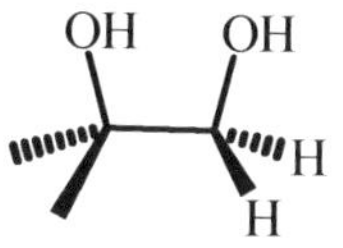

OH OH
H
H

Product B

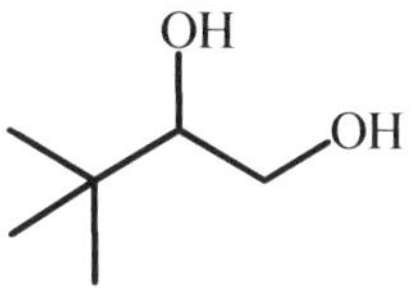

OH
OH

Product C

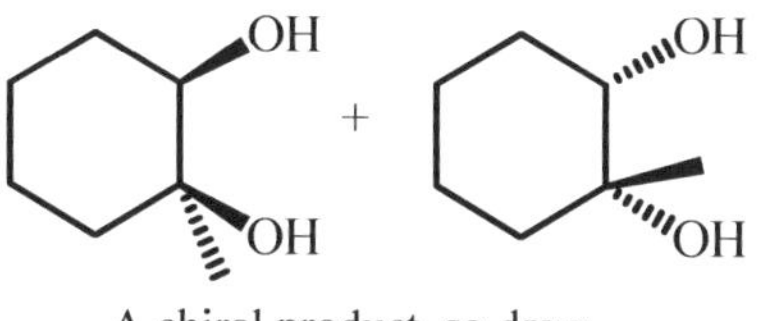

OH
OH
+
OH
OH

A chiral product, so draw isomers resulting from addition to each face

10. Ozonolysis of alkenes

a) General Net Reaction (reducing condition workup):

$$\text{H}_3\text{C}(\text{H})\text{C=C}(\text{R})\text{CH}_3 \xrightarrow[\text{2. reducing conditions}]{\text{1. O}_3\text{, low temperature}} \text{H}_3\text{C}(\text{H})\text{C=O} + \text{O=C}(\text{R})\text{CH}_3$$

Aldehyde

b) General Net Reaction (oxidizing condition workup):

$$\text{H}_3\text{C}(\text{H})\text{C=C}(\text{R})\text{CH}_3 \xrightarrow[\text{2. oxidizing conditions}]{\text{1. O}_3\text{, low temperature}} \text{H}_3\text{C}(\text{HO})\text{C=O} + \text{O=C}(\text{R})\text{CH}_3$$

Carboxylic acid

c) Notable Points

- In sharp contrast to the reactions shown so far in this section, this is a cleavage of the alkene (breakage of the C=C bond) rather than an addition.
- If a doubly-bound carbon has two non-H substituents, that fragment becomes a ketone
- If there is a H substituent on a doubly-bound carbon:
  - The H remains an H under reducing workup conditions (an aldehyde is one of the products)
  - The H becomes an OH under oxidizing workup conditions (a carboxylic acid is one of the products)
- Typical reducing workup conditions:
  - R_2S
  - Zn, H_2O
- Typical oxidizing workup conditions:
  - H_2O_2

d) Section Covered in Text:

e) Date(s) Discussed in Class:

f) Arrow-Pushing Mechanism

i) Reaction of alkene with ozone and immediate rearrangement to ozonide:

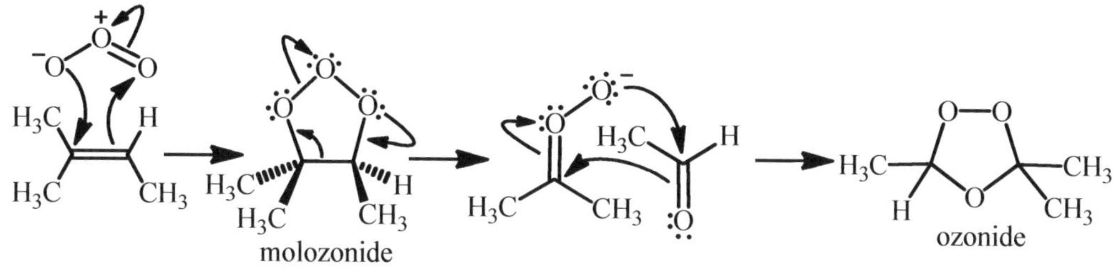

molozonide

ozonide

ii) Result of different workup conditions (no arrow-pushing):

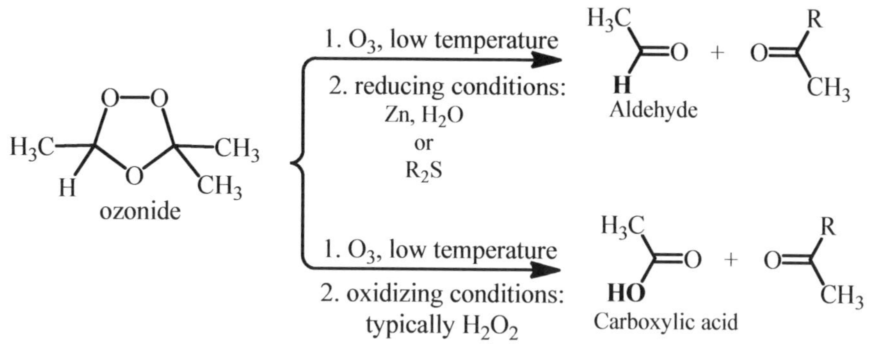

ozonide

1. O_3, low temperature

2. reducing conditions:
Zn, H_2O
or
R_2S

Aldehyde

1. O_3, low temperature

2. oxidizing conditions:
typically H_2O_2

Carboxylic acid

* Also Never Covered *

g) Self Test

i) Draw the major product(s) of each reaction:

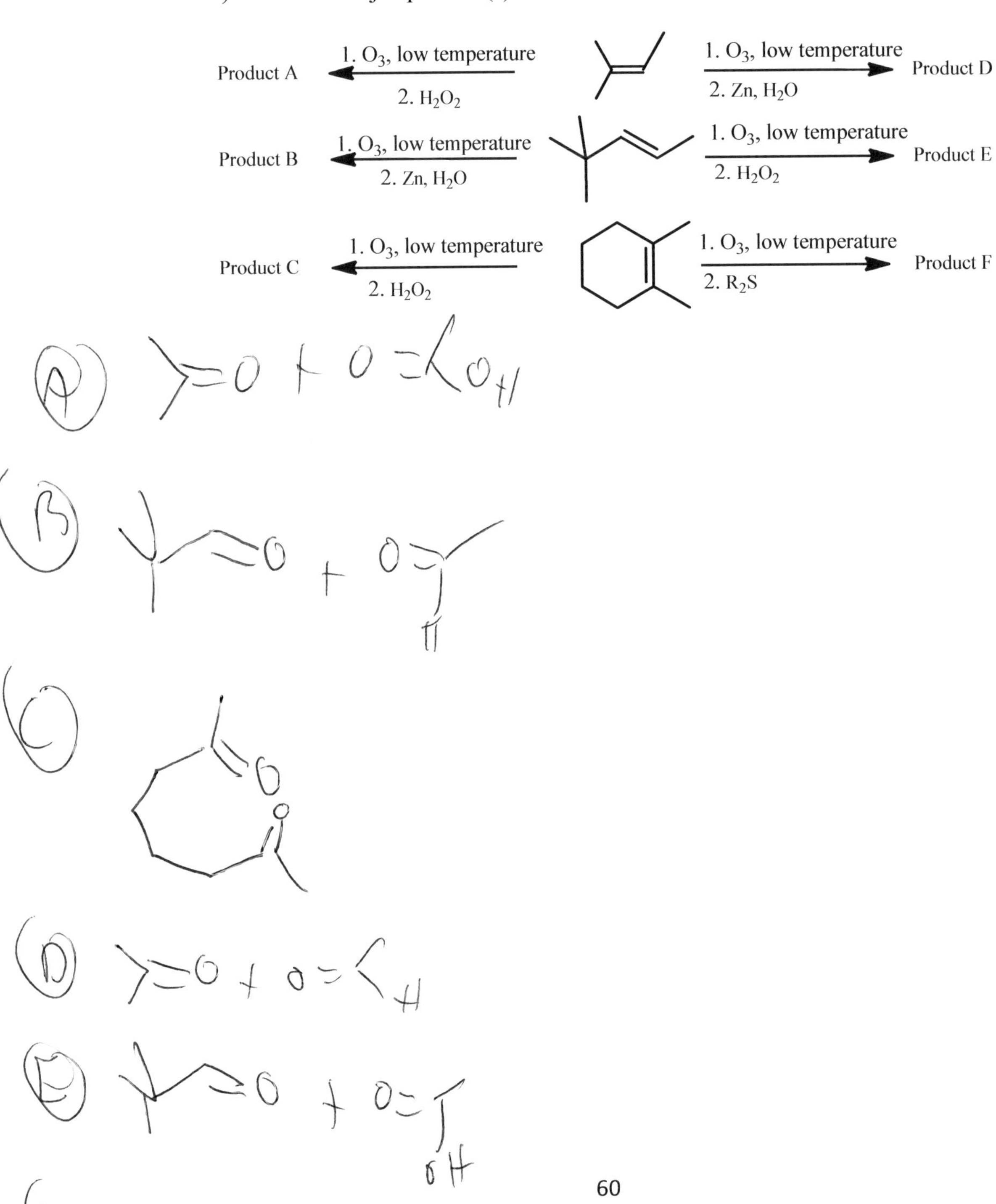

h) Answers to Self Test

i) Major products:

Product A

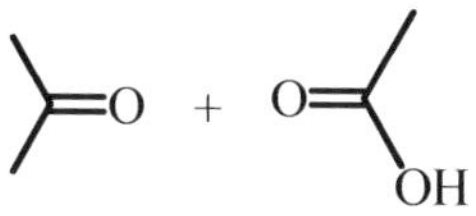

Product D

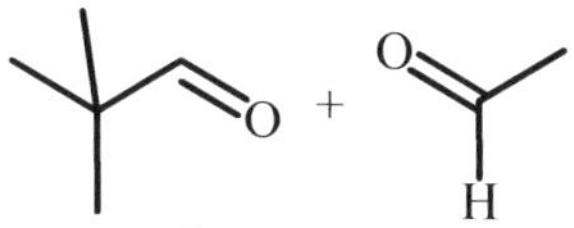

Product B

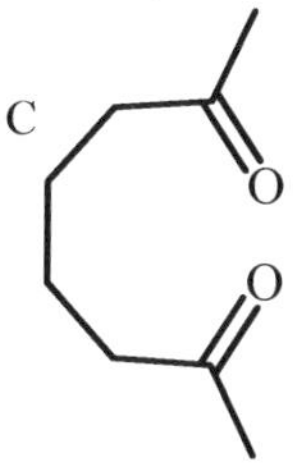

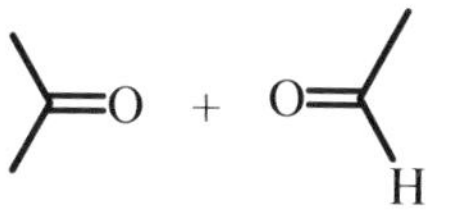

Product E

Product C

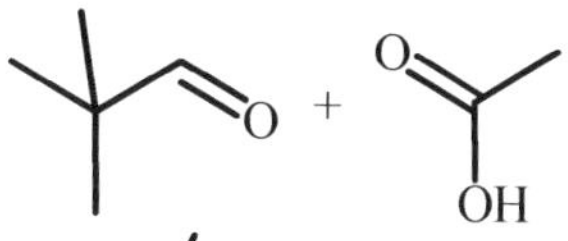

Product F

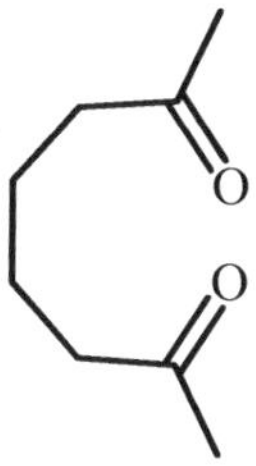

11. Reactions of alkenes involving radicals or having nontrivial mechanisms

11.1 Hydrogenation of alkenes

a) General Net Reaction using H_2:

H H

H_2

Metal Catalyst
(often Pt or Pd)

b) General Net Reaction with deuterium labeling:

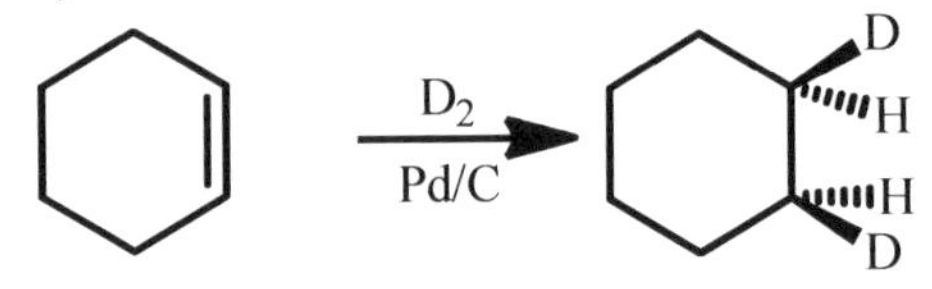

c) Notable Points

- Both groups added are the same, so you do not have to differentiate between Markovnikov or anti-Markovnikov addition
- Produces the *syn-* addition product
- No carbocation intermediate is involved
- The carbon skeleton does not rearrange during reaction
- One can use deuterium gas (D_2) in place of H_2 in this reaction; deuterium (atomic symbol D) is an isotope of H having a neutron in the nucleus
- The catalyst is often used as carbon with the metal adsorbed to the surface, so it is often listed as "Pd/C" for "palladium on carbon" or as "Pt/C" for "platinum on carbon"
- An excess of hydrogen is always used, so if there are multiple alkene units in a molecule, all of them get hydrogenated
- Benzene rings do not get hydrogenated under these conditions

d) Section Covered in Text:

e) Date(s) Discussed in Class:

f) Self Test

i) Is net addition of H_2 Markovnikov or anti-Markovnikov? Neither

ii) Is addition of H_2 predominantly *syn-*, *anti-*, or an essentially equal mixture of both?

syn

iii) Is there an intermediate that can lead to a rearrangement of the carbon atom connectivity during the reaction?

NO

iv) Draw the major product(s):

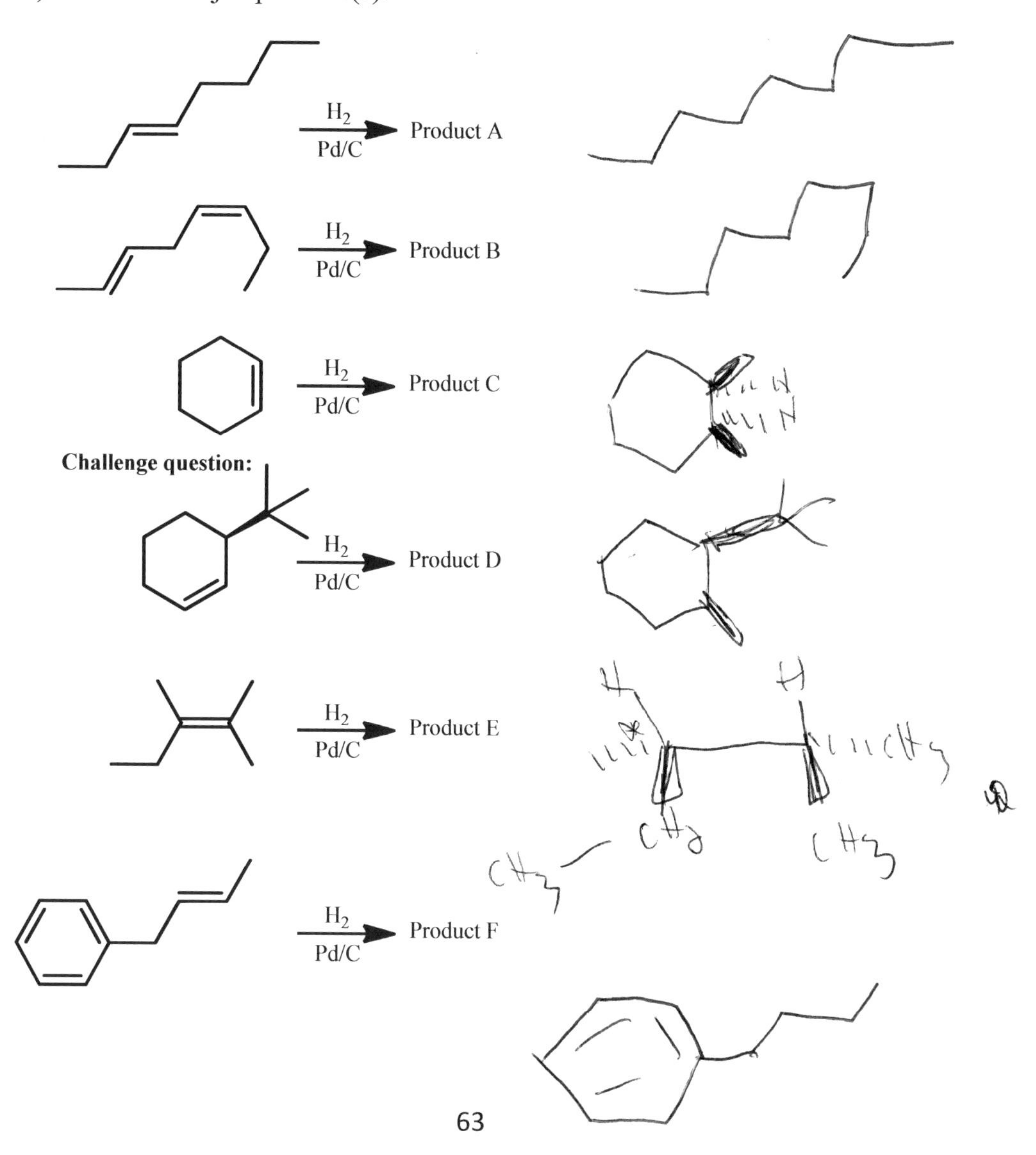

g) Answers to Self Test

i) This is a trick question! Since both groups are the same, it is not Markovnikov or anti-Markovnikov.

ii) Specifically *syn*-.

iii) No

iv) Major product(s):

Product A

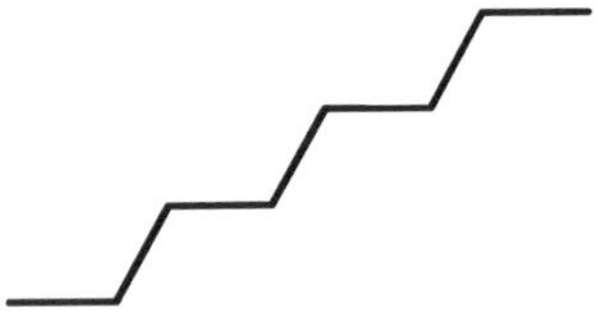

Product B

Product C (Same as:)

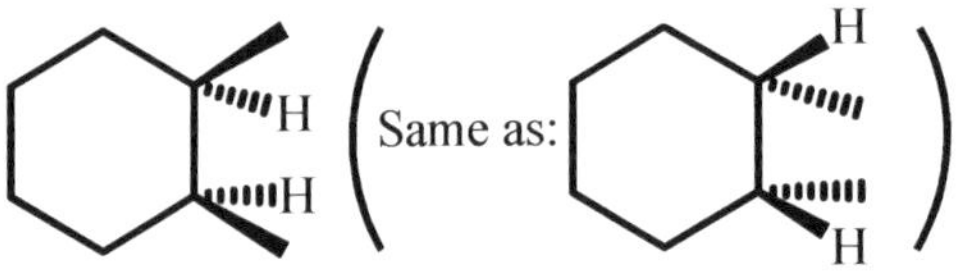

Product D

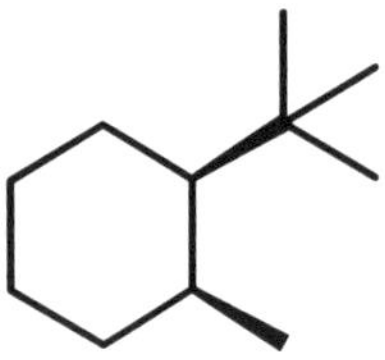

The bulky t-butyl group blocks that face from being hydrogenated at the catalyst surface, so H atoms must add to the other side!

Product E

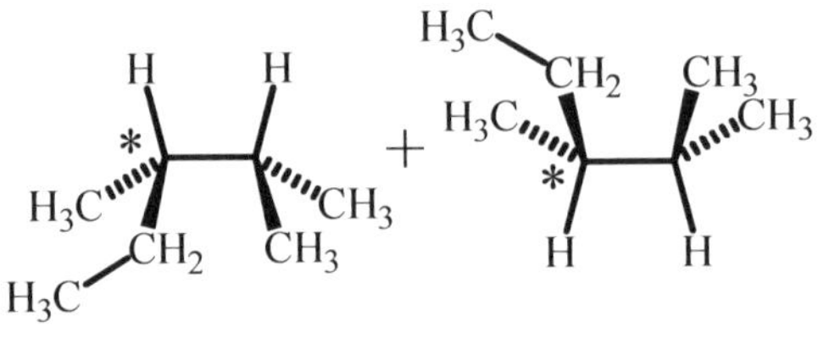

Product has a chiral center and starting materials are not chiral, so equal mixture (racemic mixture) of the two enantiomers is produced.

Product F

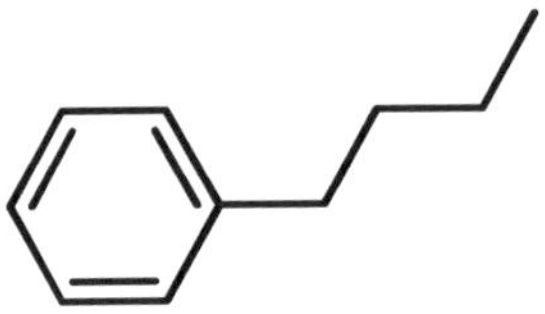

11.2 Hydrobromination of alkenes in the presence of peroxide

a) General Net Reaction:

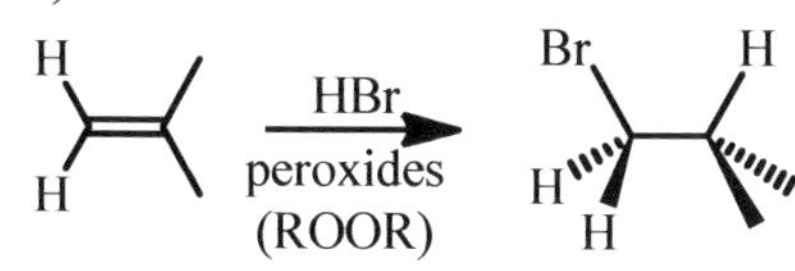

b) Notable Points

- Net anti-Markovnikov addition (-Br on the more substituted carbon)
- Essentially 1:1 mixture of *syn*- and *anti*- addition products
- A radical intermediate is involved
    - The more stable radical is the one on the more substituted carbon
    - The added H ends up where the radical forms on the substrate (see mechanism in part e)
- None of the intermediates involved can undergo a rearrangement of the carbon skeleton
- This is one of the key anti-Markovnikov addition reactions of alkenes taught in most introductory organic courses (see also Section II, reaction 8: Hydroboration / oxidation of alkenes)

c) Section Covered in Text:

d) Date(s) Discussed in Class:

e) Mechanism

i) Steps of the Radical Chain Mechanism

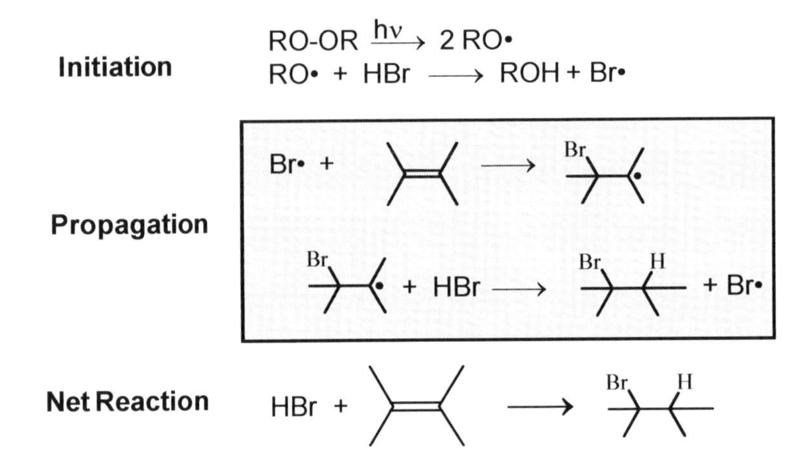

ii) Drawn as a catalytic cycle:

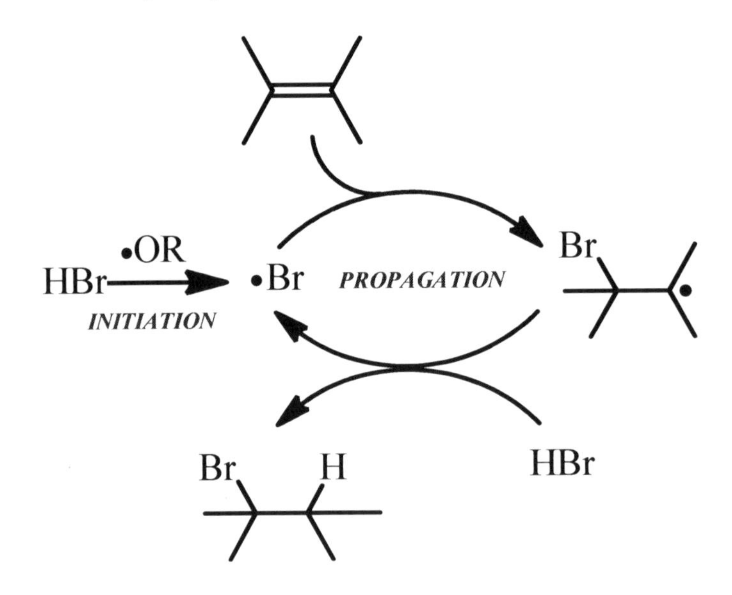

f) Self Test

i) Provide initiation, propagation and net reaction for the radical bromination of 2,3-dimethyl-2-butene using ROOR as the initiator.

not enough Room

ii) Is net addition of HBr Markovnikov or anti-Markovnikov?

Anti-Markovnikov

iii) Is addition of HX predominantly *syn-*, *anti-*, or an essentially equal mixture of both?

Equal

iv) What is the most important intermediate in this reaction (or is the reaction concerted)?

Radical intermediates

v) Can the structure of the intermediate rearrange during the reaction?

vi) Draw the major product(s) of each reaction:

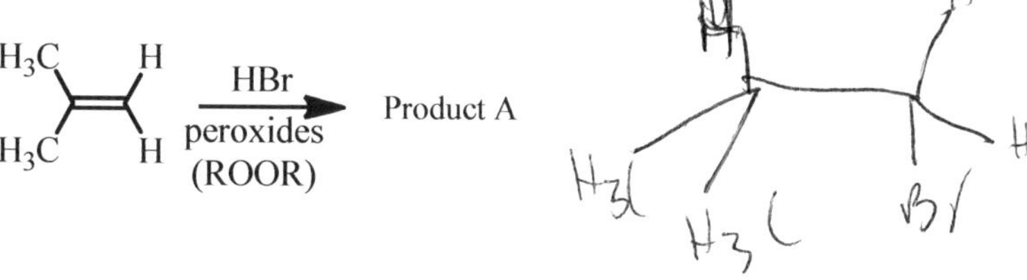

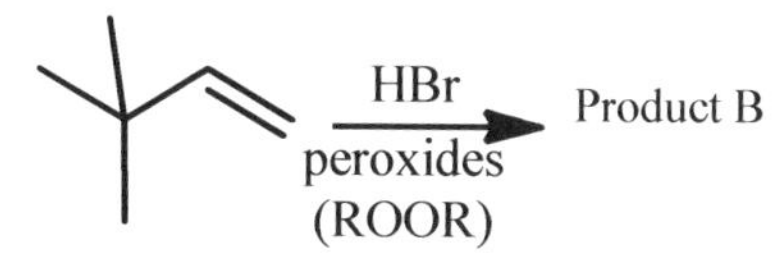

g) Answers to Self Test

i)

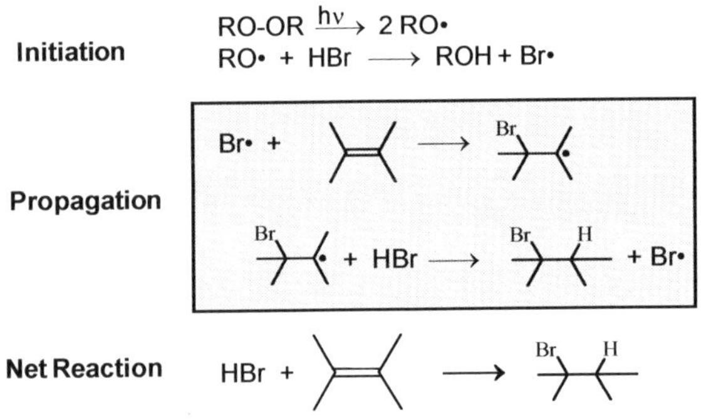

Initiation

$RO\text{-}OR \xrightarrow{h\nu} 2\,RO\bullet$

$RO\bullet + HBr \longrightarrow ROH + Br\bullet$

Propagation

$Br\bullet +$ → Br

Br + HBr → Br H + $Br\bullet$

Net Reaction

HBr + → Br H

ii) Anti-Markovnikov

iii) An essentially equal mixture of both

iv) Radical intermediates

v) No

vi) Major products:

Product A: H, Br, H_3C, H_3C, H, H

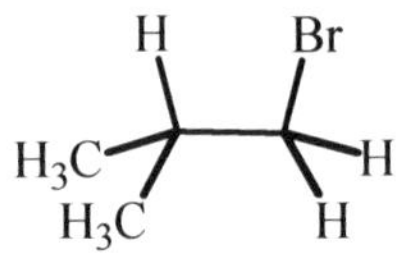

Product B: Br

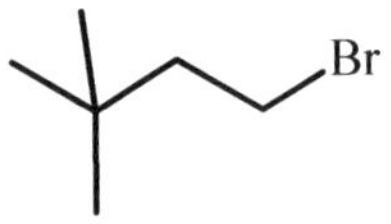

11.3 Radical halogenations at allylic or benzylic positions

a) General Net Reaction (allylic substrate):

Br

N

O O

+ hν Br + O N O

H

N-bromosuccinimide
(NBS)

b) General Net Reaction (benzylic substrate):

hν

NBS

Br + O N O

H

c) Notable Points

- This is a reaction at a site *next to* a C-C double bond, making it different from all of the previous reactions in this section.
- Allylic positions are sites next to a vinyl group (next to an alkene carbon)
- Benzylic positions are sites next to an aromatic ring, typically benzene
- Radicals are stabilized by resonance at benzylic and allylic sites
- For benzylic substrates, the halogen adds to the benzylic position, not to the aromatic ring
- For allylic substrates, the major product will typically be the one in which the halogen ends up on whichever site on the molecule has the unpaired electron in the most viable resonance contributor to the most stable radical
- In order to predict allylic substitution products, you must consider all resonance contributors to each radical that can form on the basis of the mechanism

d) Section Covered in Text:

e) Date(s) Discussed in Class:

f) Mechanism

The step-by-step mechanism provided below for radical halogenations is provided for a case in which a halogen (here Br) replaces a hydrogen atom that is on an allylic carbon (a carbon that is directly attached to a doubly-bound C). The mechanism is identical for the case when a halogen (here Br) replaces a hydrogen atom that is on a benzylic carbon (a carbon directly attached to an aromatic ring). Note that the allylic radical has an additional resonance contributor, and one must consider addition of the Br to the radical site in any resonance contributor that can be drawn by movement of the pi bond. In benzylic cases, however, the Br adds only to the benzylic site and not to the benzene ring.

Initiation $Br_2 \xrightarrow{h\nu} 2Br\cdot$

Propagation

Br• + ⟶ • + HBr

Br N O O +HBr ⟶ H N O O $+Br_2$

• + Br_2 ⟶ Br + Br•

Net Reaction:

+ Br N O O ⟶ Br + H N O O

f) Self Test

i) Provide initiation, propagation and net reaction for the radical bromination of propene using NBS and light as an initiator. Be sure to provide the structure of NBS in your answer. No Room!

ii) What is the most important intermediate in this reaction (or is the reaction concerted)? Radical intermediate

iii) Can the structure of the intermediate rearrange during the reaction? Yes

iv) Draw the major product(s) of each reaction:

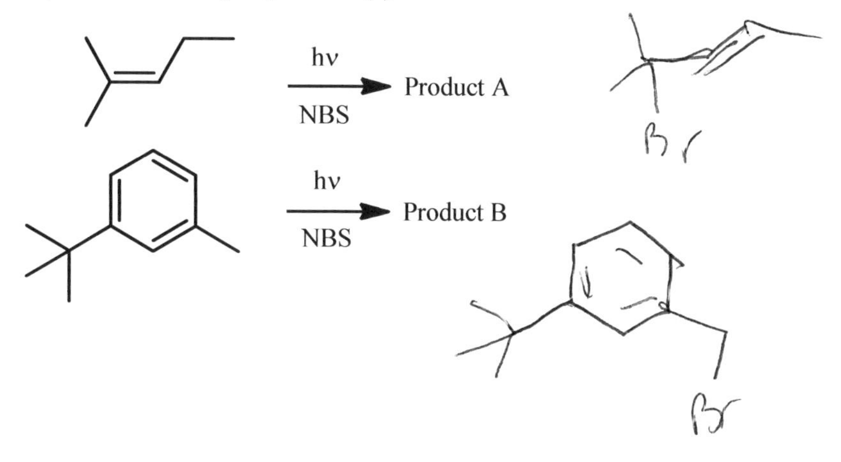

g) Answers to Self Test

i)

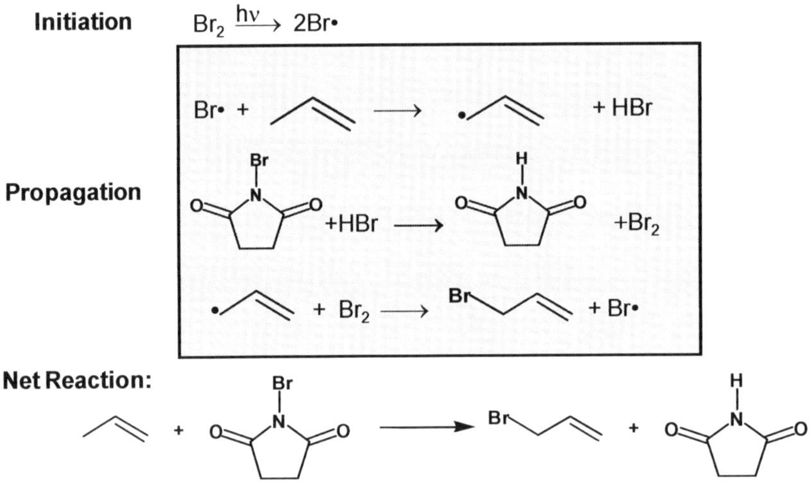

Initiation $Br_2 \xrightarrow{h\nu} 2Br\bullet$

Propagation

Br• + → • + HBr

Br N O O +HBr → H N O O +Br_2

• + Br_2 → Br + Br•

Net Reaction:

\+ Br N O O → Br + H N O O

ii) Radical intermediates are involved

iii) Yes, one can draw resonance structures of allylic radicals and all resonance contributors must be considered in determining the product. The major product comes from the best resonance contributor.

iv) Major product(s):

Product A

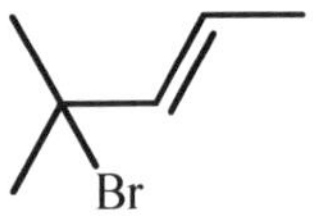

Br

Product B

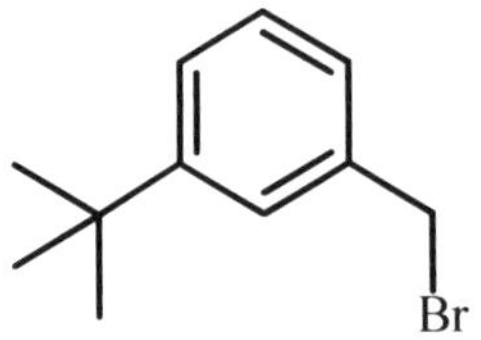

Br

12. Direct (1,2-) and conjugate (1,4-) addition of HX and X_2 to conjugated dienes

a) General Net Reaction for HX Addition:

H—X

1,2-addition (direct addition)

1,4-addition (conjugate addition)

b) General Net Reaction for X_2 Addition:

X—X

1,2-addition (direct addition)

1,4-addition (conjugate addition)

c) Notable Points

- When 1,2-addition (direct addition) occurs, the reaction at the alkene appears to be the same as if the second alkene was not beside it.
- When 1,4-addition (conjugate addition) occurs, The double bond in the product is now *in between* where the two double bonds were in the starting material
- To choose which of the two addition types will occur
    - At **low temperature** you get the **kinetic** product (product where the nucleophile (X^-) adds to the most stable resonance contributor of the **most stable cation**
    - At **high temperature** you get the **thermodynamic** product, which is the **most substituted alkene** (Zaitsev's rule)

d) Section Covered in Text:

e) Date(s) Discussed in Class

f) Arrow-Pushing Mechanism

i) Mechanism for HX addition (here, X = Br)

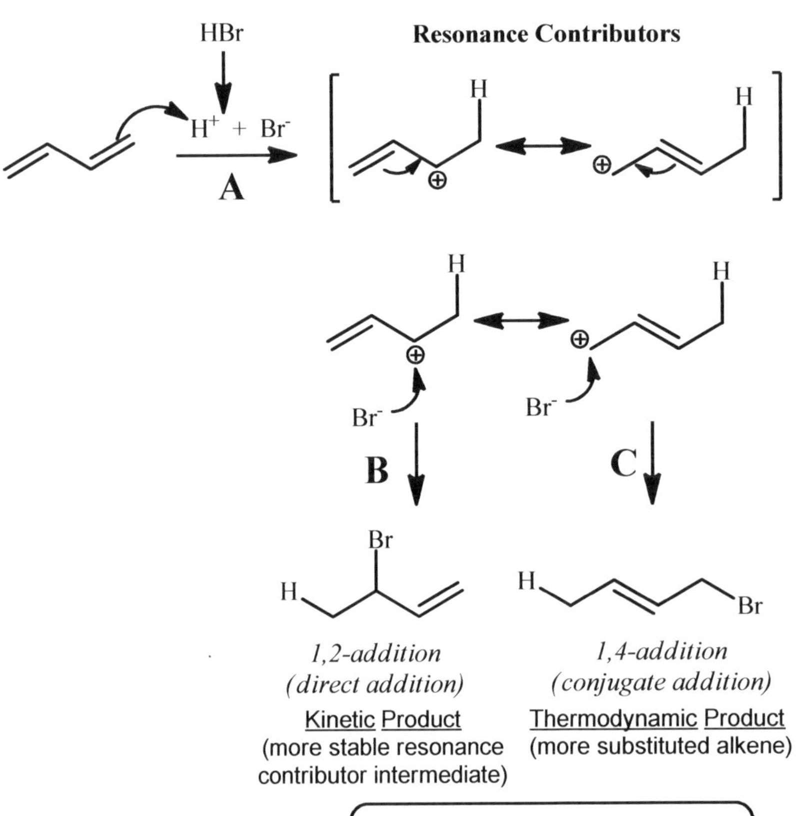

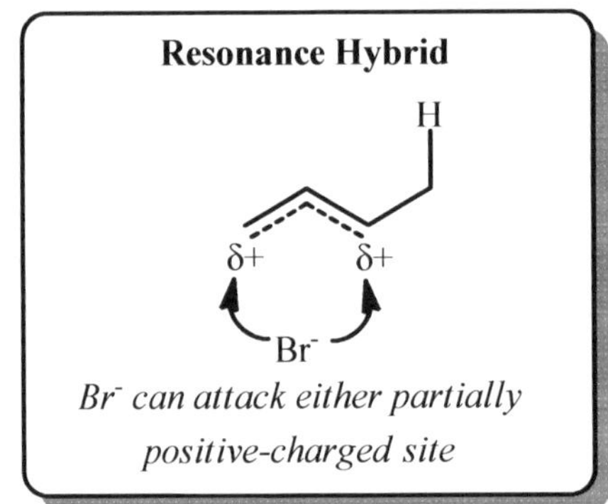

ii) Mechanism for X_2 addition (here, X = Br)

Resonance Contributors

1,2-addition *(direct addition)*	*1,4-addition* *(conjugate addition)*
<u>Kinetic</u> <u>Product</u> (more stable resonance contributor intermediate)	<u>Thermodynamic</u> <u>Product</u> (more substituted alkene)

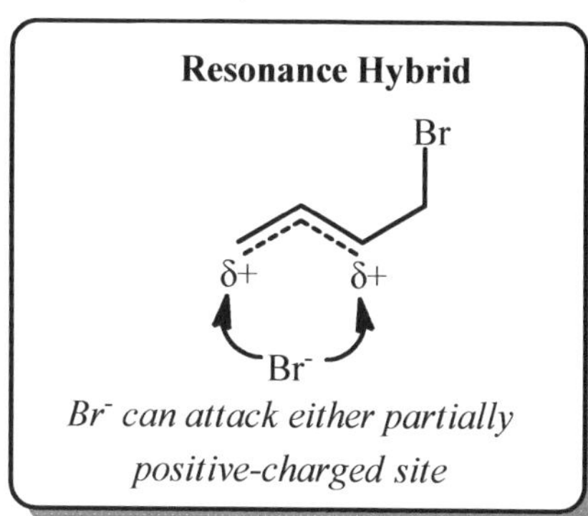

iii) Mechanism when the diene double bonds have different substituents

If there is a choice between two nonidentical alkenes in a diene, chose to have the electrophile attack the alkene that gives the more substituted cation (here, choice 2).

Choice 1: Attack the alkene on the right

2°-allyl 1°-allyl

Choice 2: Attack the alkene on the left

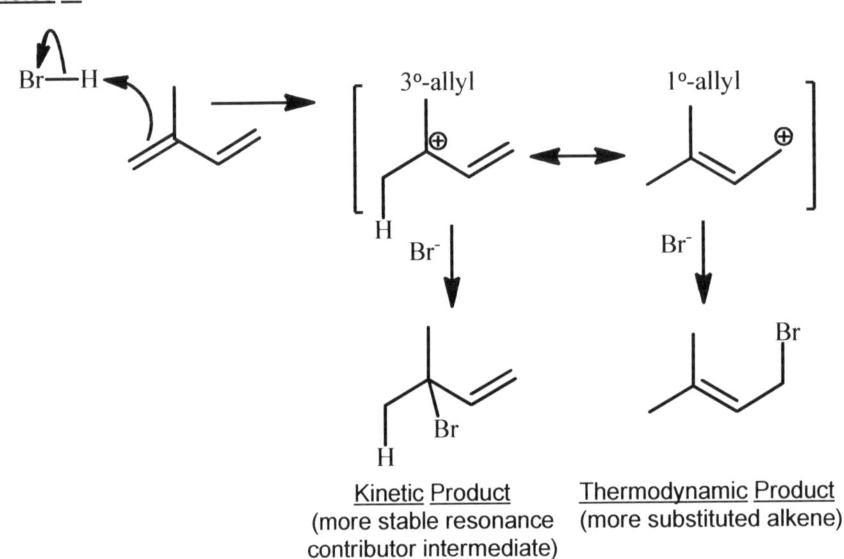

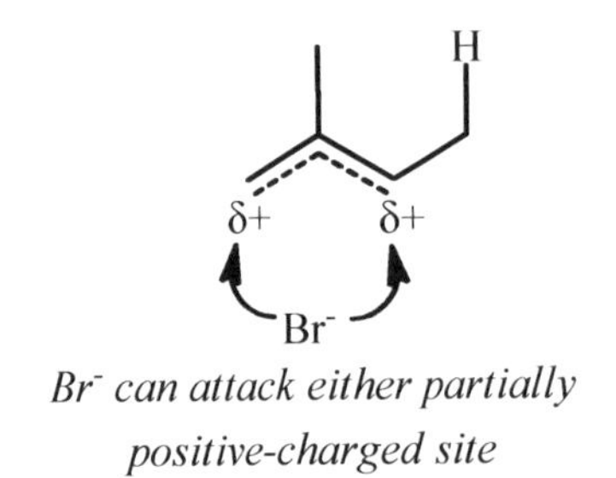

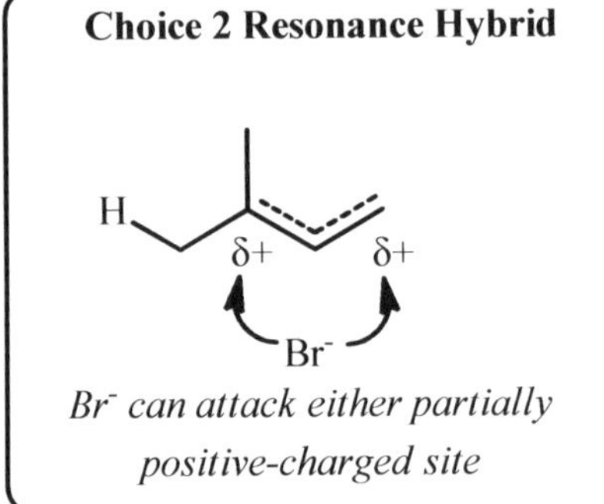

g) Self Test

i) How does one determine what the kinetic product is in terms of intermediate and product stability? most stable resonance contributor of most stable intermediate

ii) How does one determine what the thermodynamic product is in terms of intermediate and product stability? most stable product

iii) Does the kinetic product tend to be favored more at high or low temperature?

iv) Does the thermodynamic product tend to be favored more at high or low temperature?

v) Draw the major product(s):

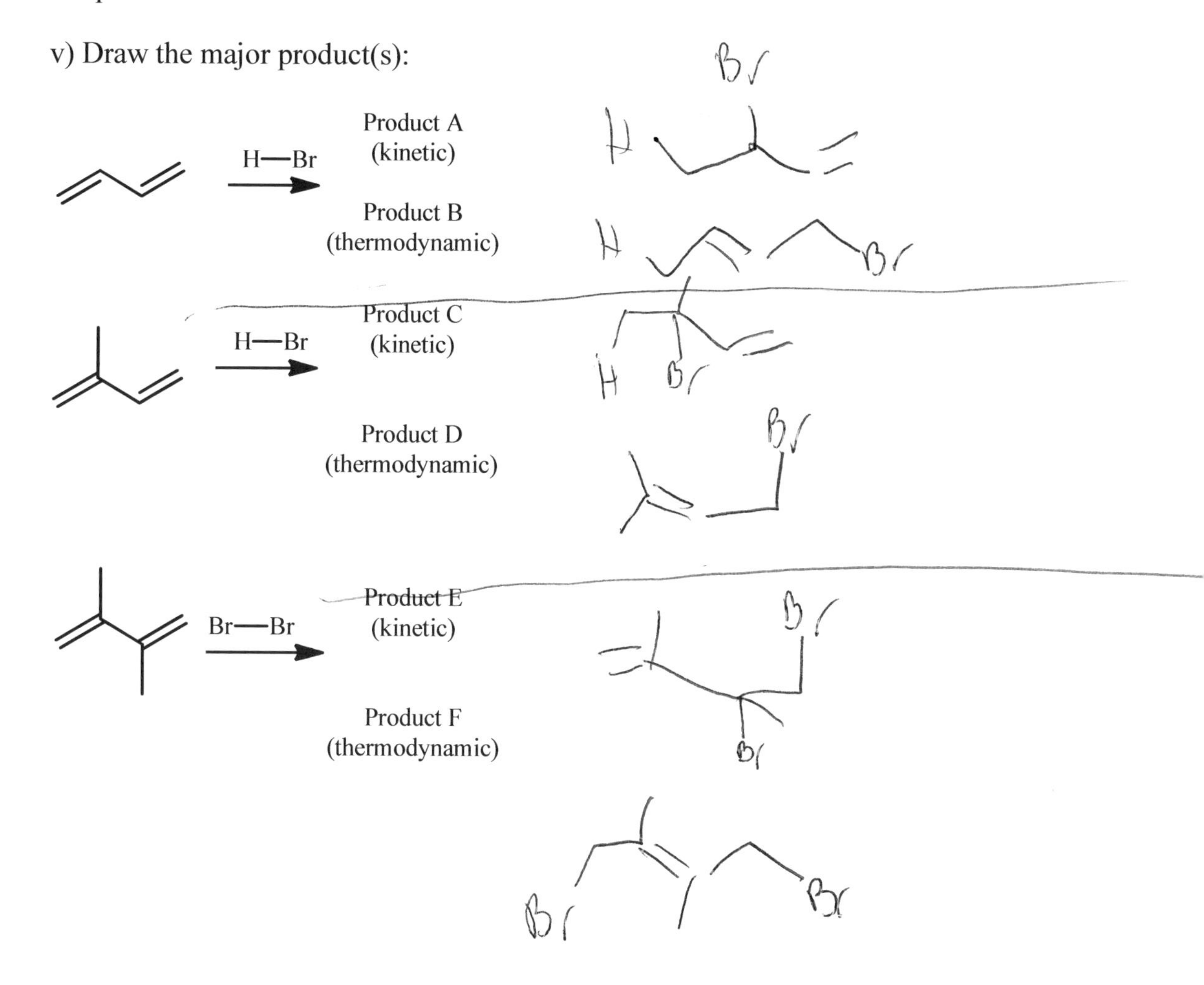

h) Answers to Self Test

i) The kinetic product derives from the most stable resonance contributor of the most stable intermediate.

ii) The thermodynamic product is simply the most stable product.

iii) low temperature

iv) high temperature

iv) Major product(s):

Product A
(kinetic)

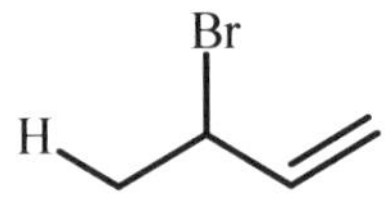

Product B
(thermodynamic)

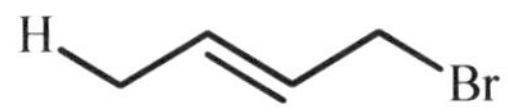

Product C
(kinetic)

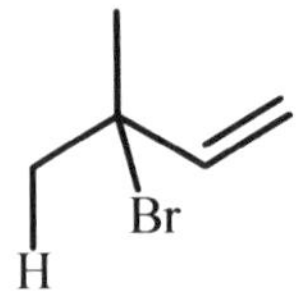

Product D
(thermodynamic)

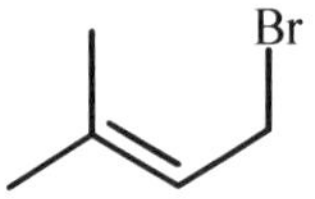

Product E
(kinetic)

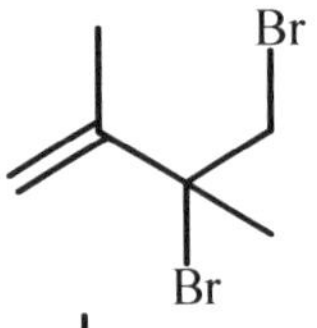

Product F
(thermodynamic)

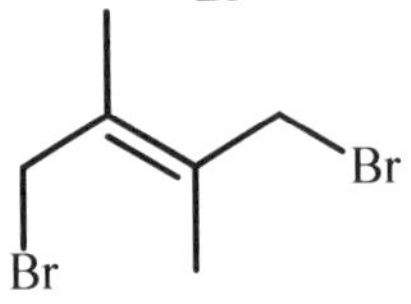

13. Diels-Alder Cycloaddition

a) General Net Reaction with Alkene as Dienophile:

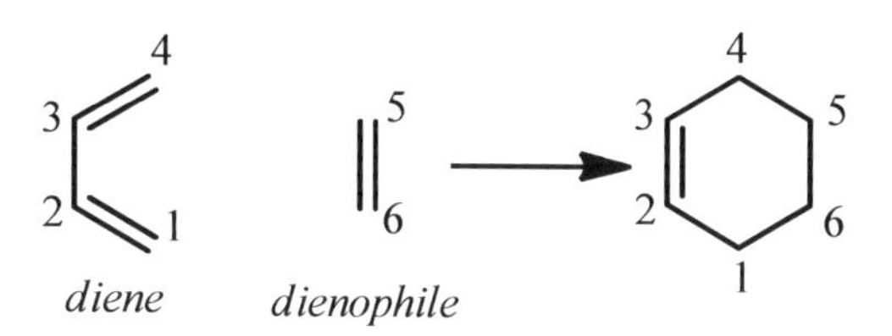

b) General Reaction with Alkyne as Dienophile:

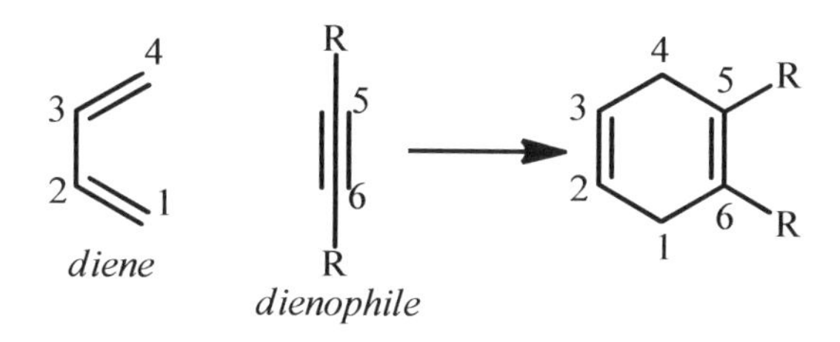

c) Notable Points

- Reaction is concerted
- Reaction involves reaction of one pi bond on a **dienophile** with a conjugated **diene** to form a new six-membered ring
- Reaction is faster when there are electron-withdrawing groups on the dienophile
- An alkene or alkyne may act as the dienophile
- The **endo** product is typically favored (see Mechanism notes in part f)

d) Section Covered in Text:

e) Date(s) Discussed in Class

f) Arrow-Pushing Mechanism

i) Simple alkene and alkyne dienophiles

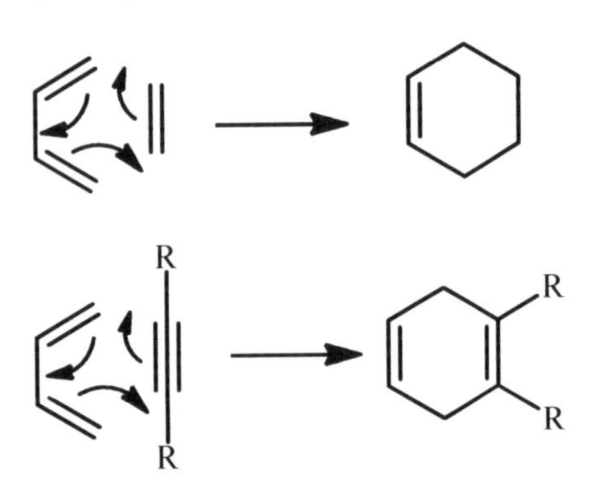

The reaction is concerted, so there are no intermediates. In a concerted reaction, it is often helpful to keep the **transition state** in mind:

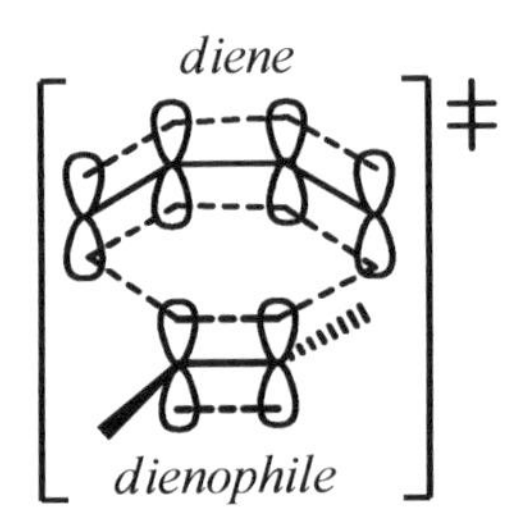

ii) Stereochemistry Note: Reaction at different faces of the alkene

Because the Diels-Alder reaction is a concerted reaction, the alkene substituents **cannot** rearrange with respect to one another during the reaction. Two important ramifications result from this observation. First, alkene substituents on a diene will remain in the same relative orientation in the product as they display in the starting material. In the reaction below, the methyl groups on the dienophile are *cis*- to each other in the starting material and also in the product, for example.

The second ramification is that when chiral centers are produced upon reaction, the *face* of the dienophile that interacts with the diene in the transition state will influence which stereoisomeric product is made. Furthermore, if neither starting material (diene or dienophile) is chiral, there will not be a preference for one of two enantiomers, so both would be produced in a 1:1 ratio. Consider the reaction of a simple *trans*-dienophile with a diene. Each of the two enantiomers derives from interaction of the diene with a specific *face* of the alkene in the transition state:

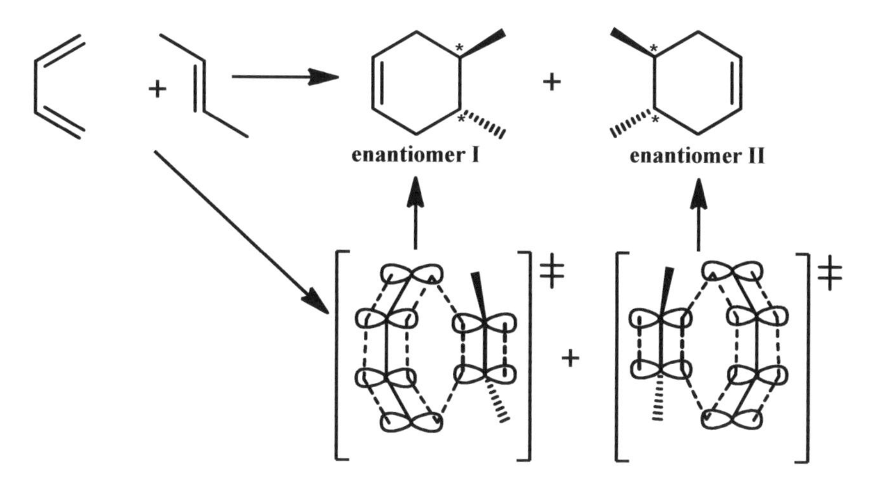

iii) Example using a cyclic diene:

iv) Stereochemistry note: *exo*- versus *endo*-product

When you have a cyclic diene, you sometimes have a choice between two products that differ only in the position of the new pi bond in the product relative to the substituent that came from the dienophile. In the **endo product**, the substituent points towards the same direction as the pi bond. In the **exo product**, the substituent points away from the pi bond. The rule is that **if the substituent R has a pi bond in it itself, the endo product is the major product**.

endo product:
R is down and the pi bond in the product is down

exo product:
R is up and the pi bond in the product is down

g) Self Test

i) What is the most important intermediate in this reaction (or is the reaction concerted)?

ii) Draw the major product(s) of each reaction:

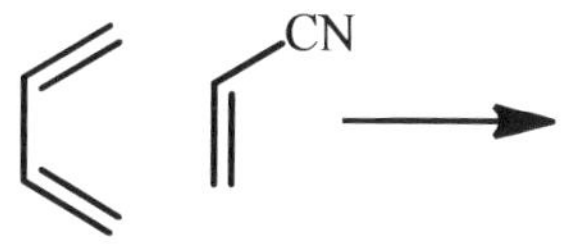

CN

Product A

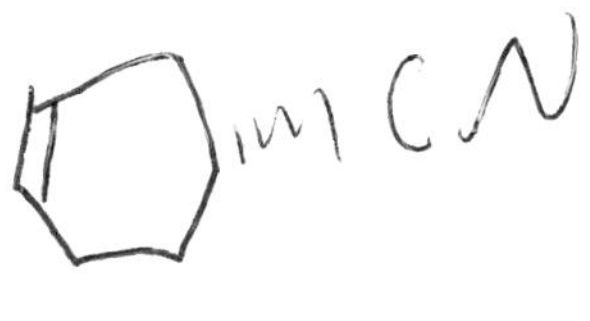

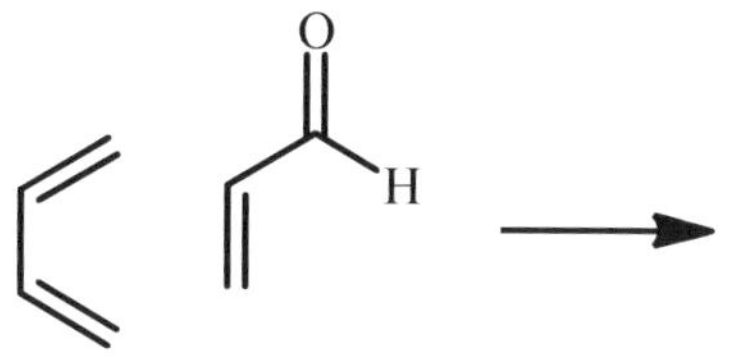

O

H

Product B

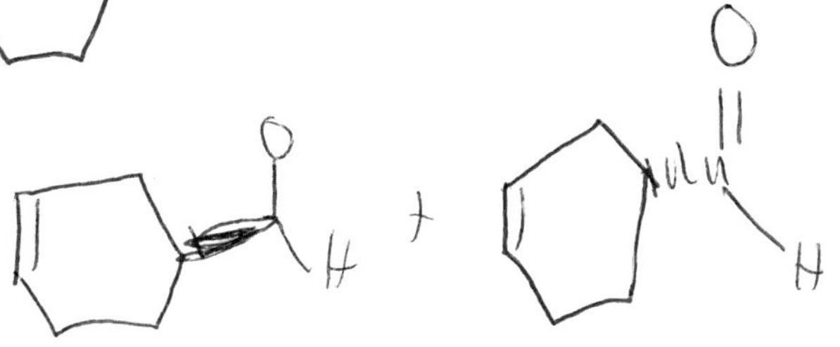

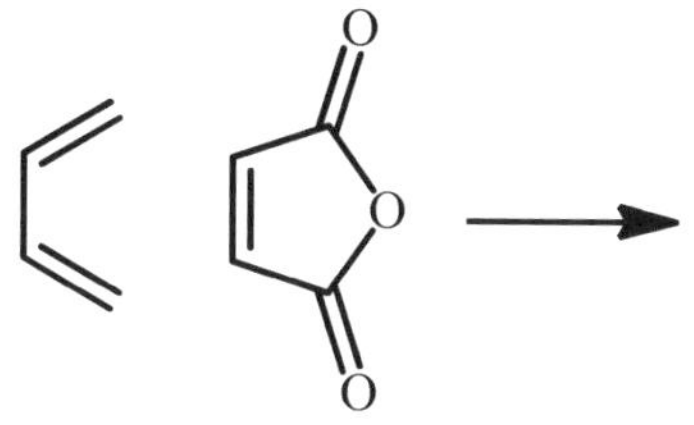

O

O

O

Product C

O OCH_3

H_3CO O

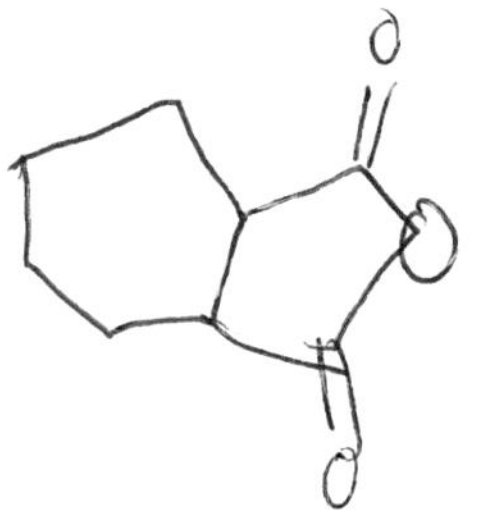

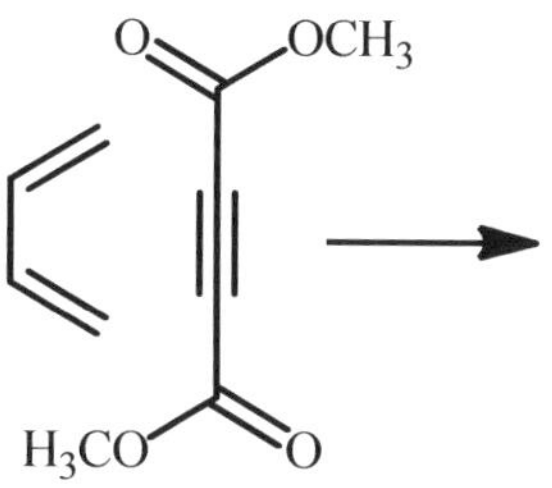

Product D

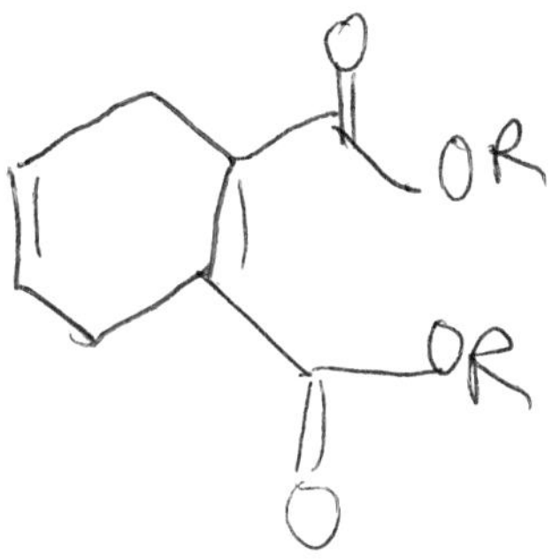

h) Answers to Self Test

i) This reaction is concerted

ii) Major product(s):

Product A

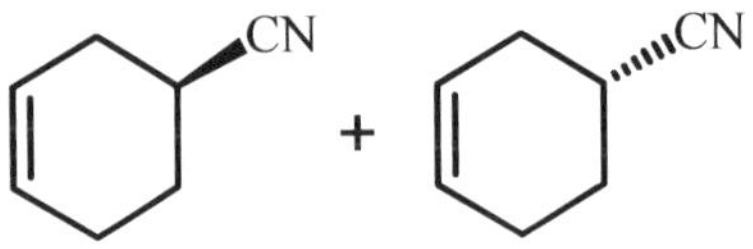

Product B

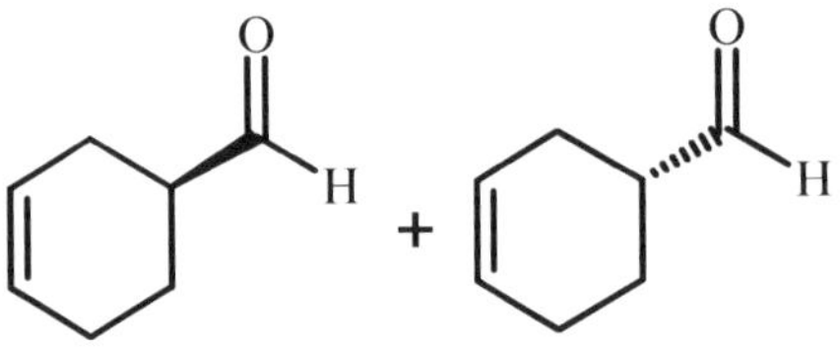

Product C

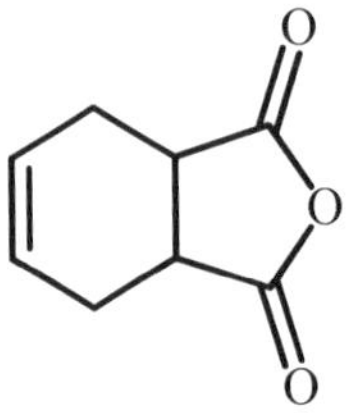

Product D

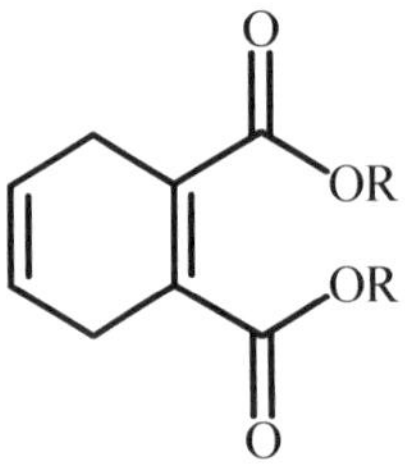

III. Reactions of Alkynes

1. Hydrohalogenation of Alkynes with HX (X = Cl, Br, I)

a) General Net Reaction:

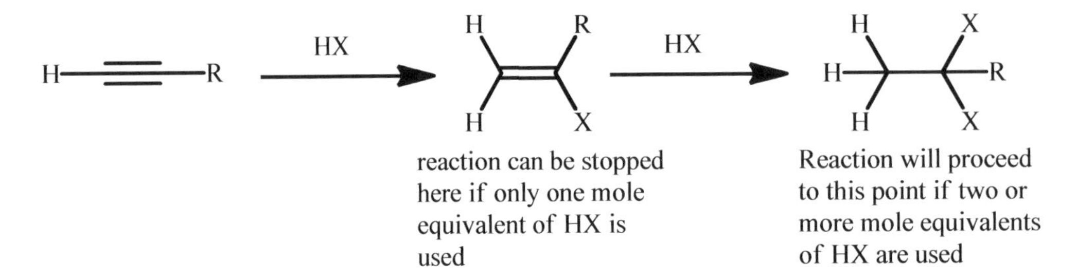

b) Notable Points

- For the first reaction of HX with the **alkyne**:
  - Markovnikov addition (X on the more substituted carbon)
  - The *anti*- addition product is usually favored
  - A pi complex intermediate is involved
  - The pi complex intermediate cannot rearrange
  - Stronger acids react faster (HI > HBr > HCl)

- For the second reaction of HX with the **alkene**:
  - See Reaction II.1

c) Section Covered in Text:

d) Date(s) Discussed in Class

e) Arrow-Pushing Mechanism

Note that steps A and B show the mechanism for the addition of the first equivalent of HX. The anion attacks the more substituted side. The reaction can be stopped at that point if only one equivalent of HX is used. Steps C and D show the mechanism for the addition of another equivalent of HX (this is Reaction II.1 shown earlier in this text). Both additions are Markovnikov, with the X ending up on the more substituted carbon. Mechanistic details for the addition of HX to an alkene, including an example that also involves carbocation rearrangement, are also provided in **Reaction II.1**.

H^+ R—≡—H —A→ R—≡—H (with H^+, X^-) *π complex* —B→ (R)(X)C=C(H)(H) + H^+ —C→ (R)(X)$C^{\oplus}$—C(H)(H)(H) + X^- —D→ (X)(X)(R)C—C(H)(H)(H)

f) Self Test

i) Provide the name for each mechanistic step shown on the previous page (choices for mechanistic steps are given on page 8).

Step A = Coordination Step C = Electrophilic Addition

Step B = Nucleophilic Addition Step D = Nucleophilic Addition

ii) Is net addition of HX Markovnikov or anti-Markovnikov?

iii) Is addition of HX predominantly *syn-*, *anti-*, or an essentially equal mixture of both?

Anti

iv) What is the most important intermediate in this reaction (or is the reaction concerted)?

The pi complex

v) Can the structure of the intermediate rearrange during the reaction? NO

vi) Draw the major product(s) of each reaction:

1 equiv HBr → Product A

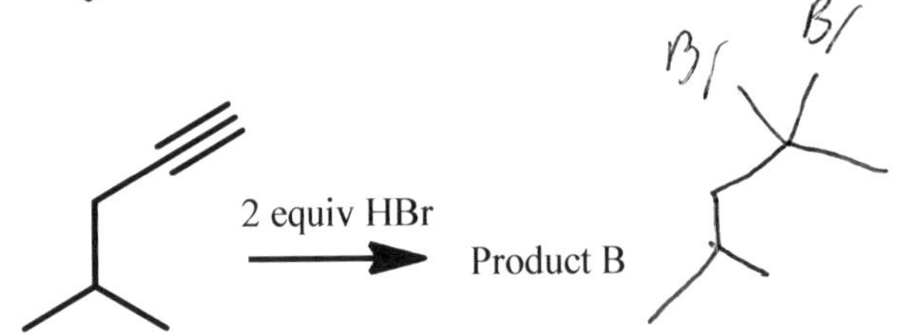

1 equiv HBr → Product C

g) Answers to Self Test

i) Provide the name for each mechanistic step shown on the previous page (choices for mechanistic steps are given on page 8).

Step A = coordination

Step B = nucleophilic addition

Step C = electrophilic addition

Step D = nucleophilic addition

ii) Markovnikov

iii) Predominantly *anti-*

iv) The pi complex

v) No

vi) Major product(s):

Br
H
H
Product A

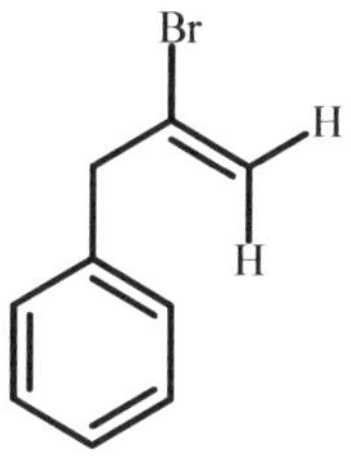

Br Br
Product B

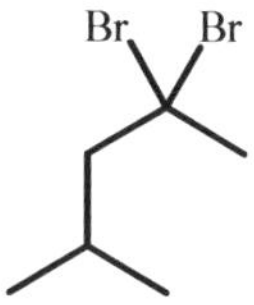

H
Br
Product C
Br
+
H

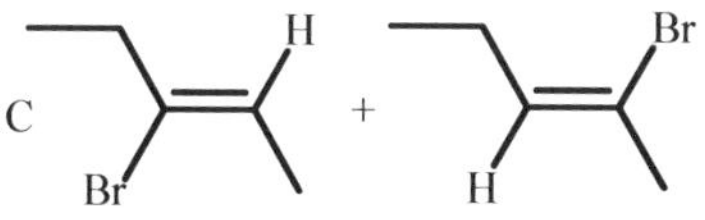

2. Halogenation of alkynes

a) General Net Reaction:

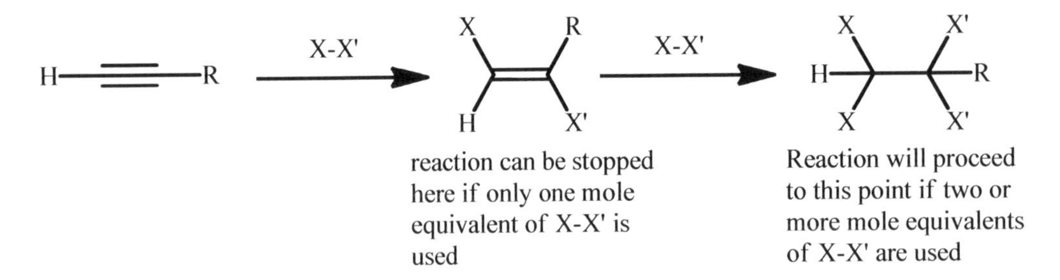

b) Notable Points (for addition of the first X-X' to the alkyne)

- Markovnikov addition if the two halogens are different. This means that the more electronegative halogen ends up on the more substituted carbon if the two halogens are different (like in Br-Cl)
- Produces the *anti-* addition product
- Carbocation intermediate is involved
- The carbocation intermediate cannot rearrange because the positive charge is on an *sp*-hybridized carbon
- Unlike halogenations of an alkene, there is no halonium intermediate involved
- If more than one equivalent of X-X' is used, the halogenations of the alkene product proceeds as for Reaction II.3

c) Section Covered in Text:

d) Date(s) Discussed in Class

e) Arrow-Pushing Mechanism

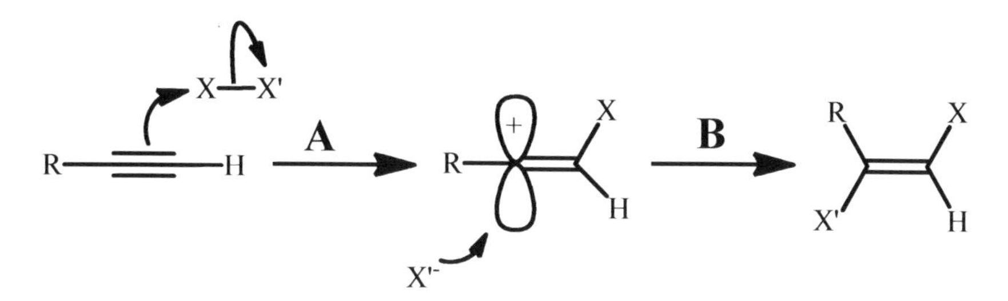

If more than one equivalent of X-X' is used, the halogenations of the alkene product proceeds as for Reaction II.3

f) Self Test

i) Is net addition of X-X' Markovnikov or anti-Markovnikov?

ii) Is addition of X-X' predominantly *syn-*, *anti-*, or an essentially equal mixture of both?

Anti

iii) What is the most important intermediate in this reaction (or is the reaction concerted)?

carbocation

iv) Can the structure of the intermediate rearrange during the reaction? NO

v) Draw the major product(s) of each reaction: what?

vi) Draw the major product(s) of each reaction:

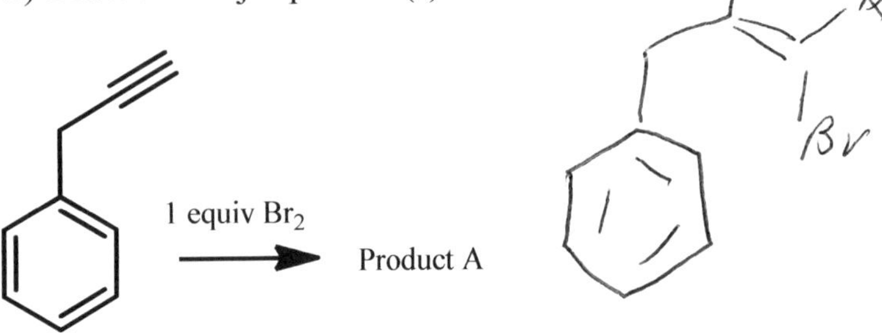

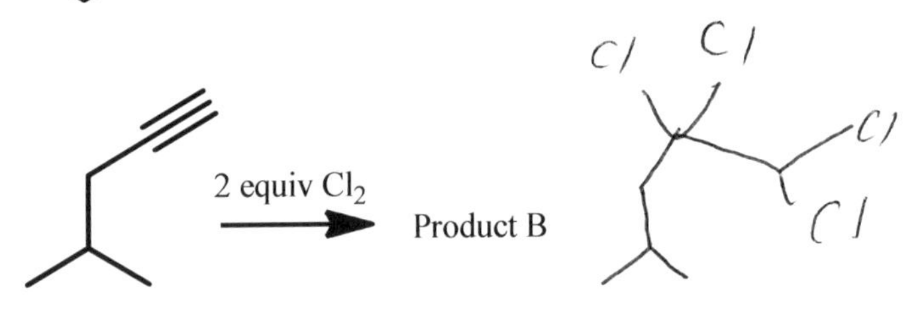

1 equiv Br-Cl → Product C

Cl Br + Br Cl

g) Answers to Self Test

i) Markovnikov (for practical purposes, this only matters when X and X' are different like in Br-Cl)

ii) Selectively *anti*-addition

iii) A carbocation intermediate is involved

iv) No, because the carbocation intermediate in this case is on an *sp*-hybridized carbon

v) Major product(s):

vi) Draw the major product(s) of each reaction:

Product A

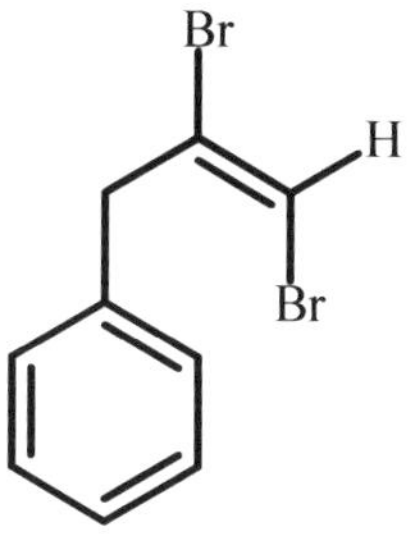

Product B

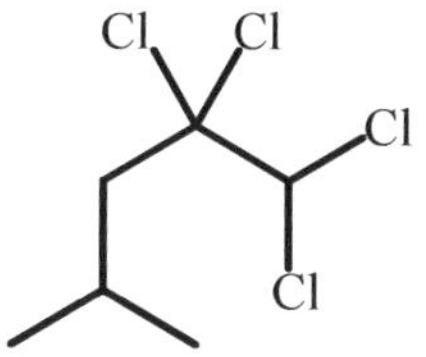

Product C +

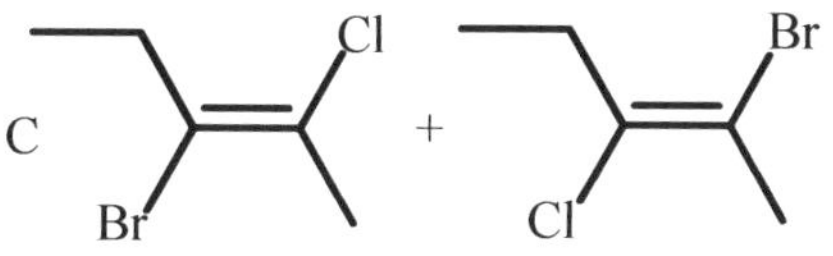

3. Tautomerization

a) General Net Reaction:

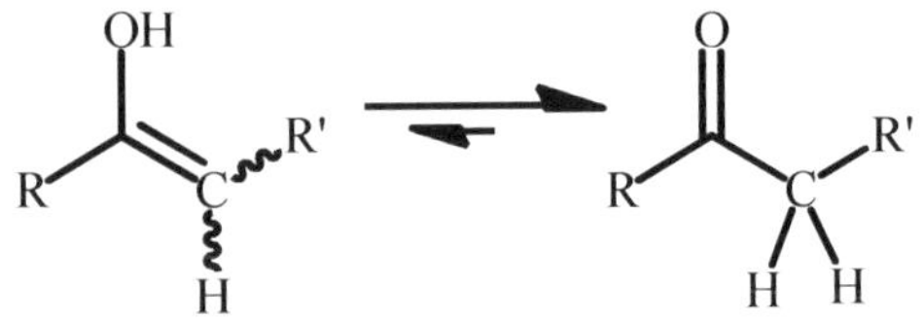

b) Notable Points

- An enol is a compound in which an OH is on a doubly-bound carbon (as in the general net reaction, left side)
- Most enols tautomerize to form the carbonyl (C=O) compound as the major isolable product form
- Look for enols as products of reactions; if they are formed, you must redraw as the carbonyl form (ketone or aldehyde).

c) Section Covered in Text:

d) Date(s) Discussed in Class

e) Arrow-Pushing Mechanism

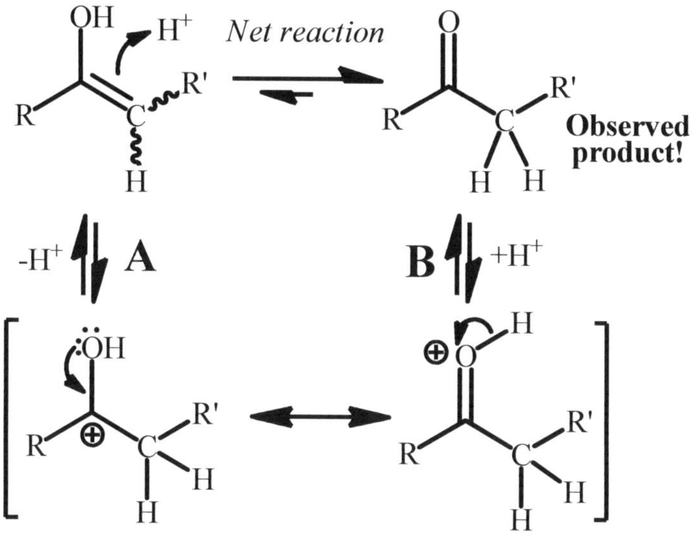

f) Self Test

i) Provide the missing structure (enol or carbonyl) for each of the following cases. If there is no tautomerization possible, write "no tautomer"

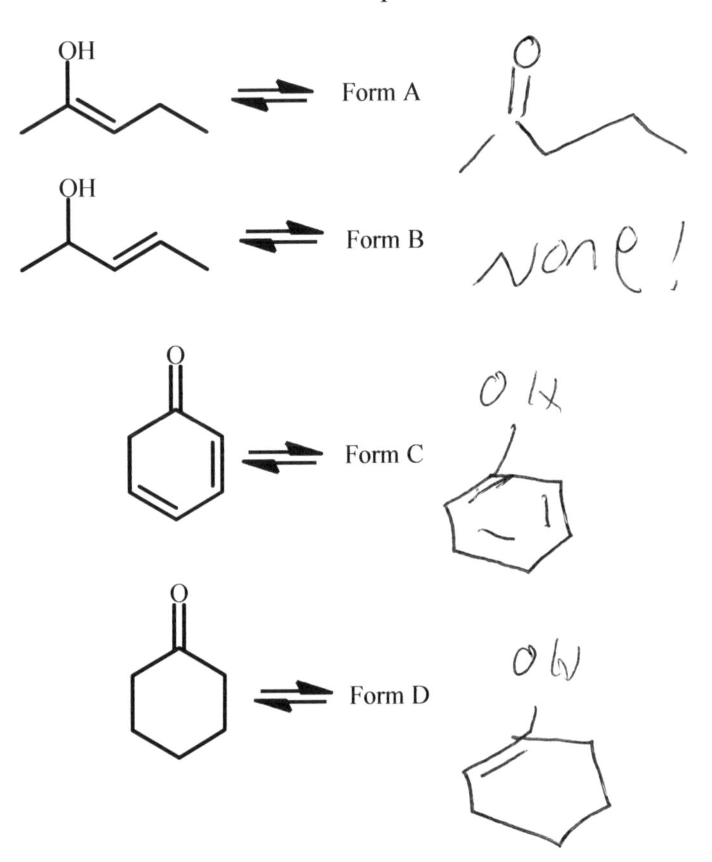

g) Answers to Self Test

i) Missing forms:

Form A

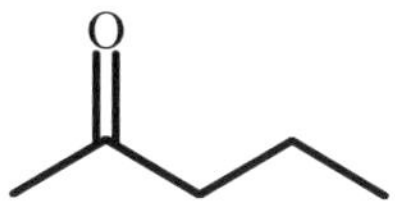

O

Form B **NO TAUTOMER; starting material is NOT an enol!**

Form C

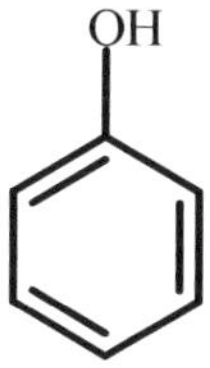

OH

Form D

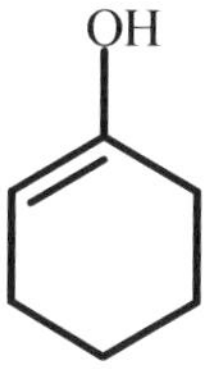

OH

4. Hydroboration / oxidation of alkynes

a) General Net Reaction:

$$R-C\equiv C-H \xrightarrow[\text{2. } OH^-,\ H_2O_2,\ H_2O]{\text{1. } BH_3/THF} R(H)_2C-C(=O)H$$

b) Notable Points

- Anti-Markovnikov addition; an –OH ends up on the **less** substituted carbon to form the enol, which tautomerizes to the carbonyl (see Section III, reaction 3: Tautomerization)
- If the alkyne starting material is terminal an aldehyde is produced as the final product
- If the alkyne starting material is internal a ketone is produced as the major product
- No carbocation intermediate is involved
- This is a three-reaction sequence:
  - 1. Hydroboration adds -H and -BR_2
  - 2. In oxidation, the –BR_2 piece is converted to –OH
  - 3. The enol so produced tautomerizes to a carbonyl compound

c) Section Covered in Text:

d) Date(s) Discussed in Class

e) Arrow-Pushing Mechanism

partial positive charge on carbon to which H adds

syn addition

tautomerize

- $[BR_2O]$-

"anti-Markovnikov Product"

Enol! Will tautomerize!

For arrow-pushing mechanism of tautomerization, see reaction III.3.

f) Self Test

i) Draw the major product(s) of each reaction:

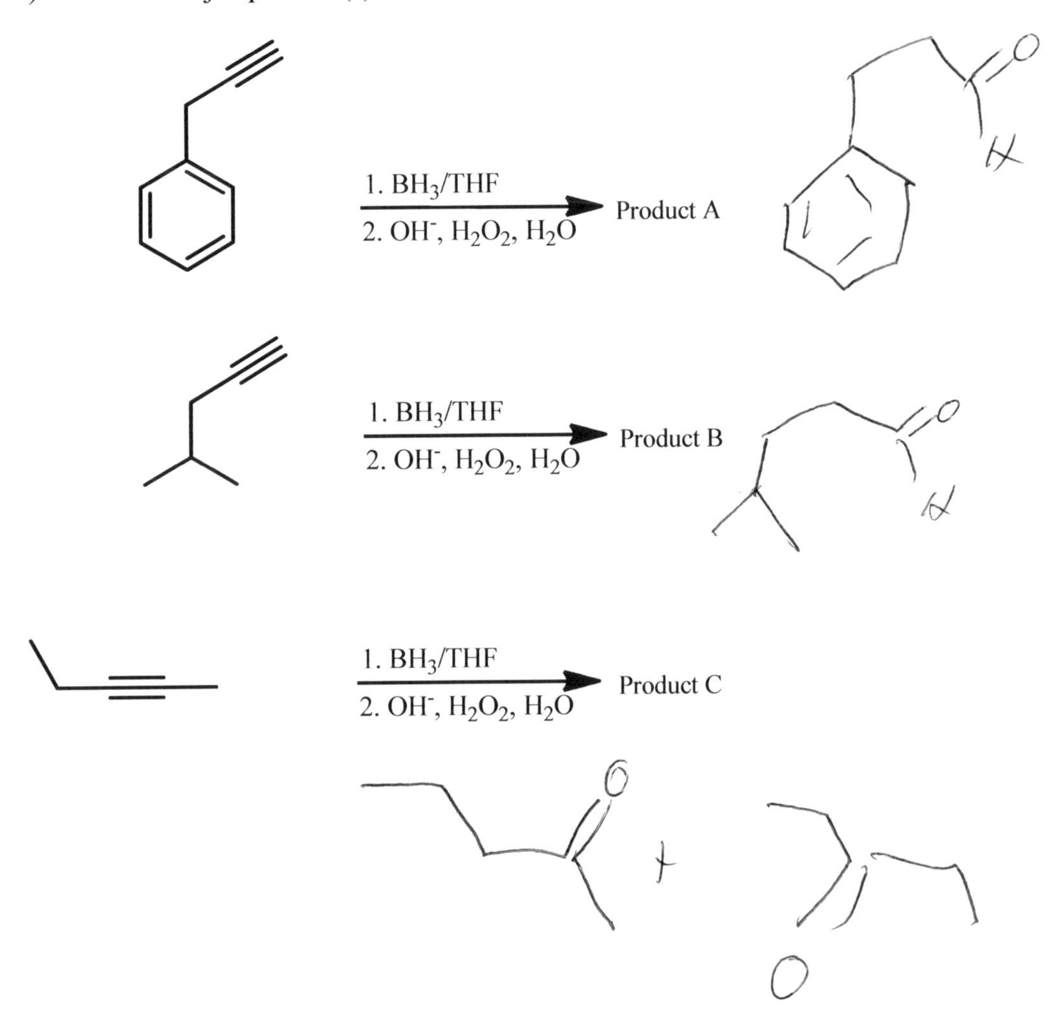

g) Answers to Self Test

i) Major product(s):

Product A

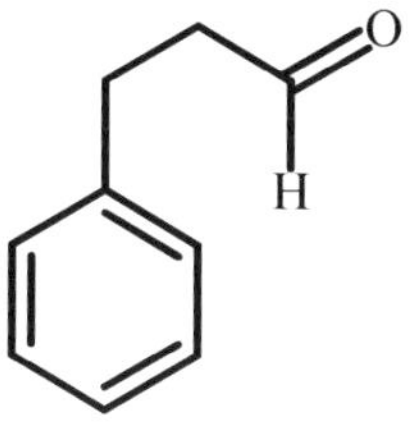

O

H

Product B

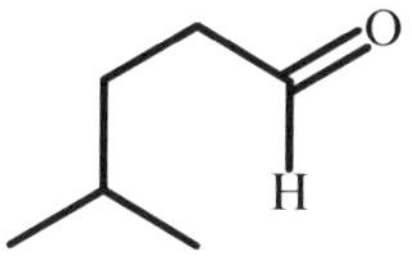

O

H

Product C

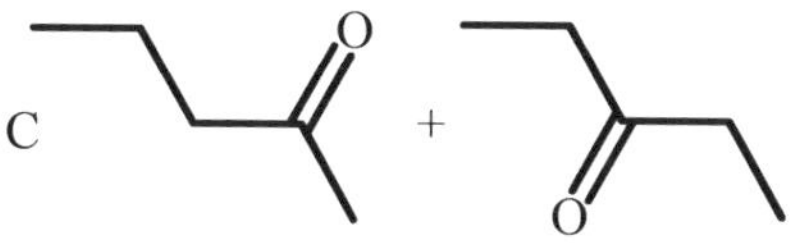

O

+

O

5. Hydration of alkynes

a) General Net Reaction:

$$R-C\equiv C-H \xrightarrow[H_2SO_4,\ H_2O]{HgSO_4} R-C(=O)-CH_3$$

b) Notable Points

- Markovnikov addition; an –OH ends up on the **less** substituted carbon to form the enol, which tautomerizes to the carbonyl (see Section III, reaction 3: Tautomerization)
- Unlike hydroboration / oxidation of alkynes, hydration yields a ketone whether the alkyne starting material is internal or terminal (only ethyne would make an aldehyde, ethanol, upon hydration)
- Unlike hydration of alkenes, no carbocation intermediate is involved in hydration of alkynes
- Unlike hydration of alkenes, hydration of alkynes requires a metal ion like Hg^{2+} to activate the alkyne to further reactions

c) Section Covered in Text:

d) Date(s) Discussed in Class

e) Arrow-Pushing Mechanism

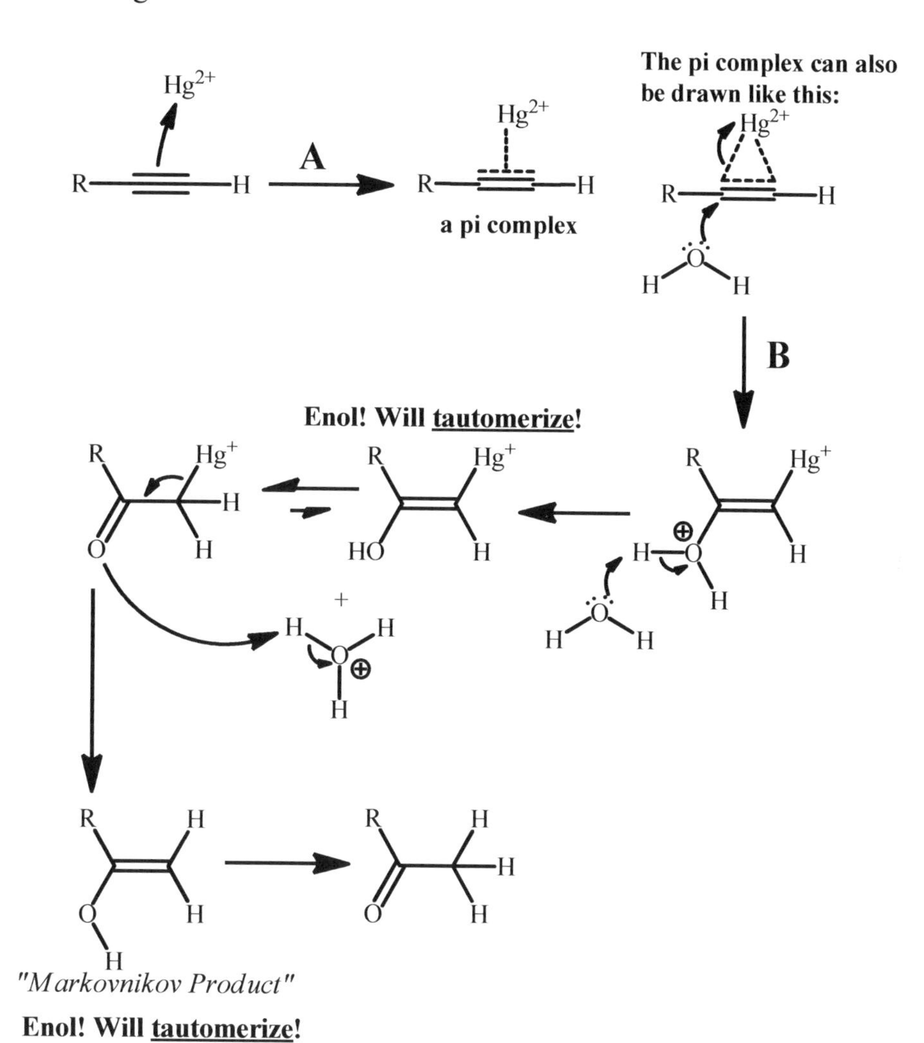

Note that step A is a simple coordination to form a **pi complex** (see basic reaction step I.1, example 5).

f) Self Test

i) Provide the name for each mechanistic step shown on the previous page (choices for mechanistic steps are given on page 8).

Step A = Coordination

Step B = Nucleophilic Addition

ii) Draw the major product(s) of each reaction:

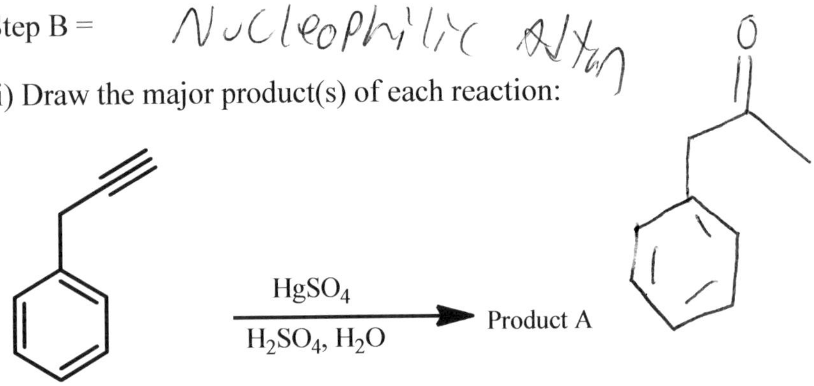

$HgSO_4$ / H_2SO_4, H_2O → Product A

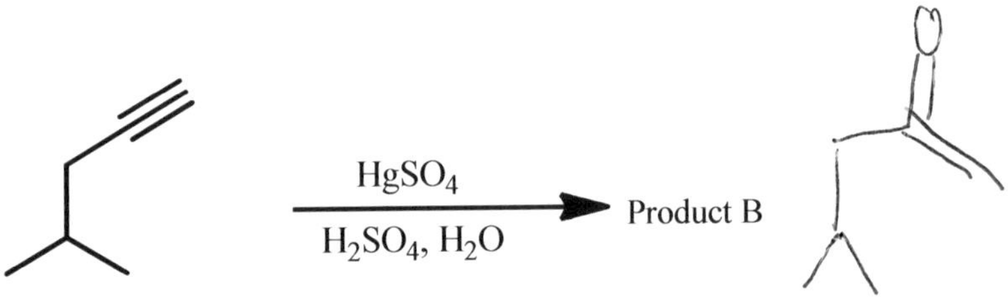

$HgSO_4$ / H_2SO_4, H_2O → Product B

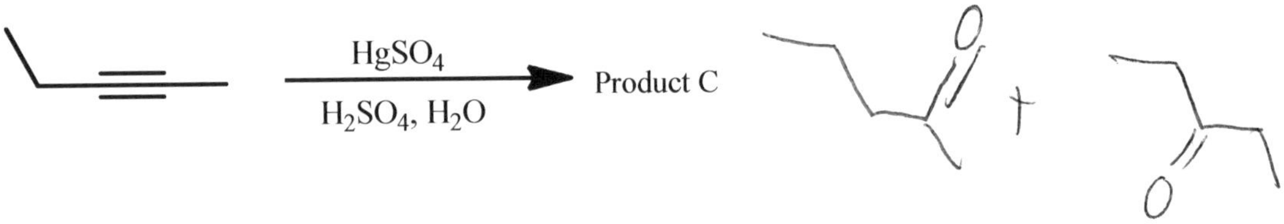

$HgSO_4$ / H_2SO_4, H_2O → Product C

g) Answers to Self Test

i) Mechanistic Steps

Step A = Coordination

Step B = Nucleophilic addition

ii) Major Product

Product A

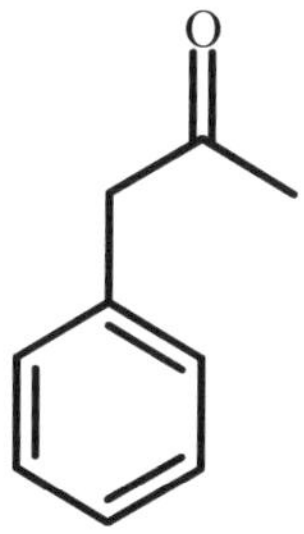

Product B

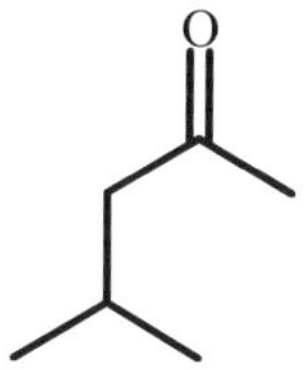

Product C +

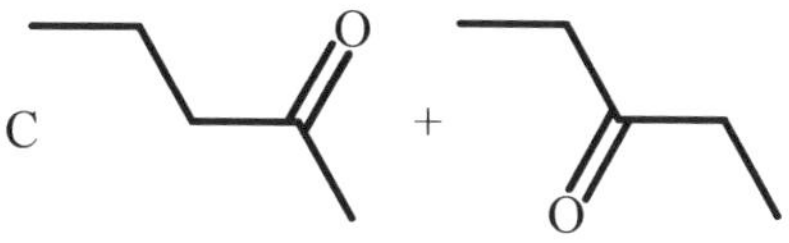

6. Hydrogenation of Alkynes using H_2

a) General Net Reaction with catalytic Pt or Pd:

$$R-C\equiv C-R \xrightarrow[\text{catalytic Pt or Pd}]{H_2}$$

Product is an **alkane**

b) General Net Reaction with Lindlar's catalyst:

$$R-C\equiv C-R \xrightarrow[\text{Lindlar's Catalyst}]{H_2}$$

Product is a *cis*-alkene

c) Notable Points

- None of these reaction is either Markovnikov or anti-Markovnikov because the group(s) added to each carbon are identical
- With H_2 and Pd/C or Pt/C, the reaction will always make an alkane; the reaction cannot be stopped after only one addition of H_2
- When Lindlar's catalyst is used, the *cis* alkene is always produced.

d) Section Covered in Text:

e) Date(s) Discussed in Class

f) Self Test

i) Draw the major product(s) of each reaction:

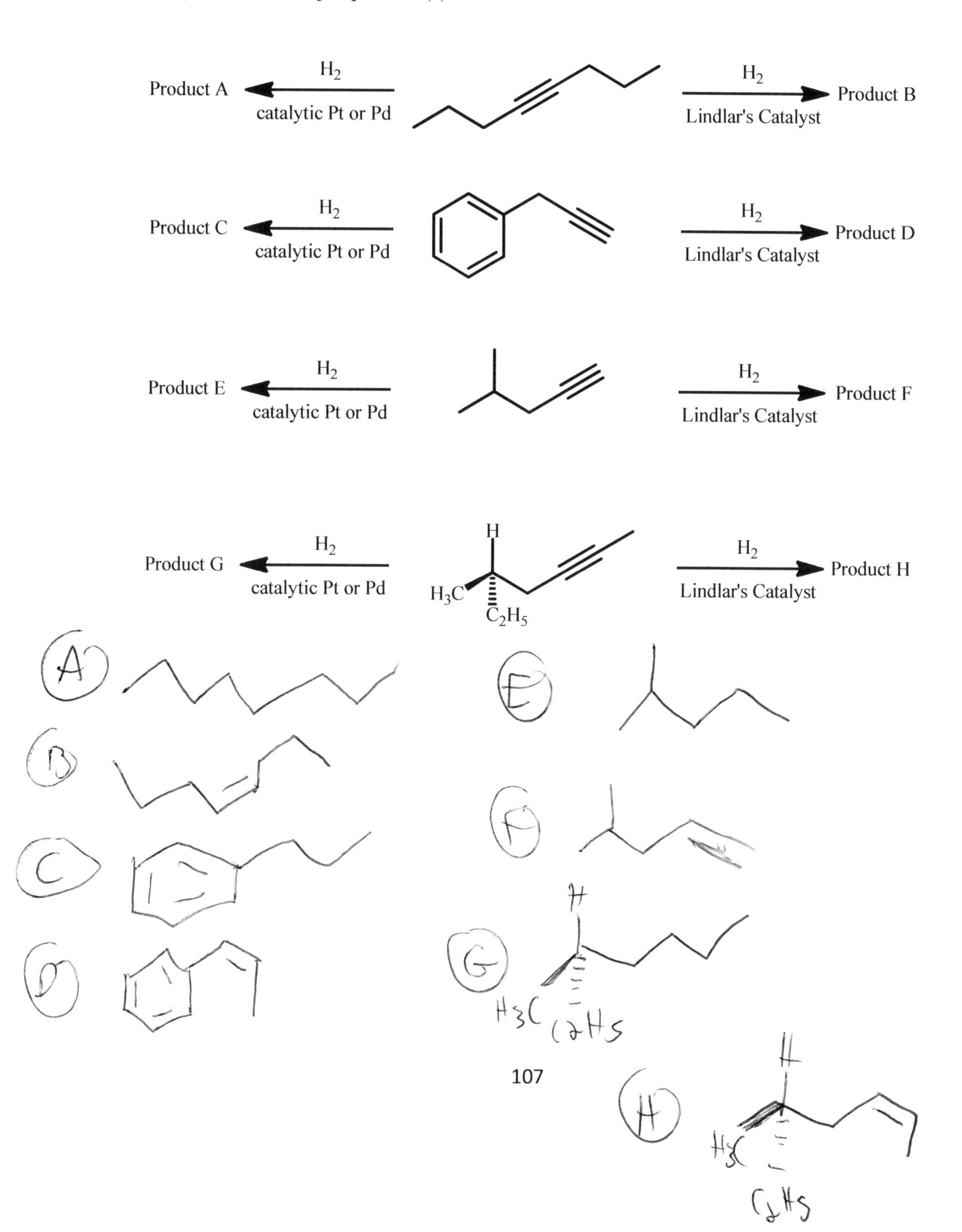

g) Answers to Self Test

i) Major product(s):

Product A

Product B

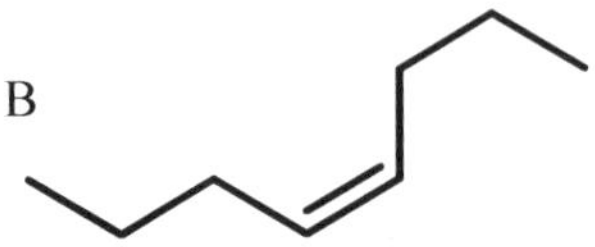

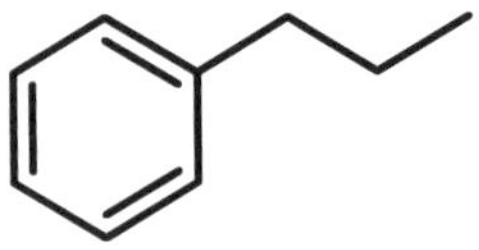

Product C

Product D

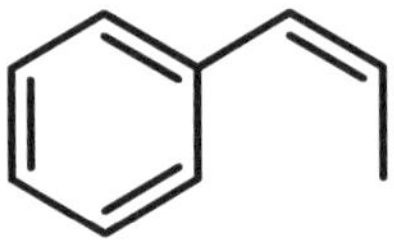

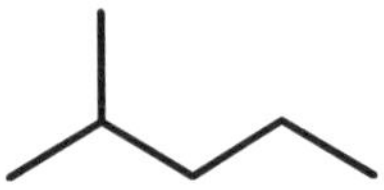

Product E

Product F

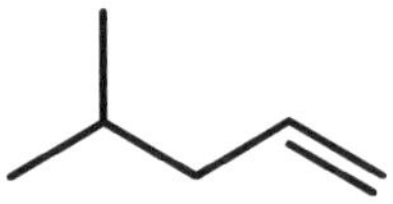

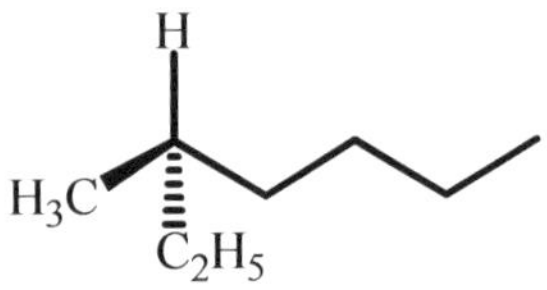

H

H_3C

C_2H_5

Product G

Product H

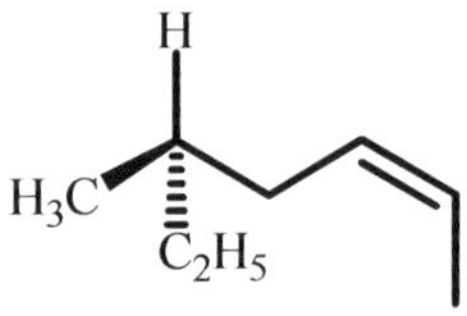

H

H_3C

C_2H_5

7. Reduction of Alkynes to *trans*-alkenes using an alkali metal

a) General Net Reaction:

R—C≡C—H $\xrightarrow[NH_3(l)]{Na\ (or\ K)}$ (R)(H)C=C(H)(R)

Product is a *trans*-alkene

b) Notable Points

- This reaction produces the *trans*-alkene as the major product
- The mechanism involved electron transfer and radical intermediates, differentiating it from the other reaction typically taught in introductory organic chemistry classes

c) Section Covered in Text:

d) Date(s) Discussed in Class

e) Arrow-Pushing Mechanism

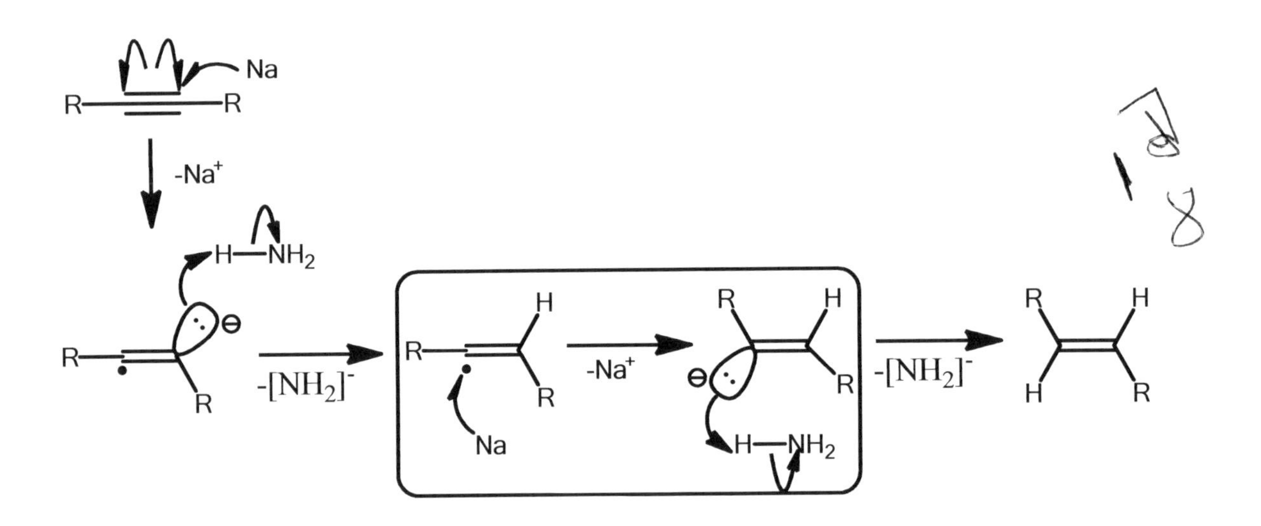

f) Self Test

i) Draw the major product(s) of each reaction:

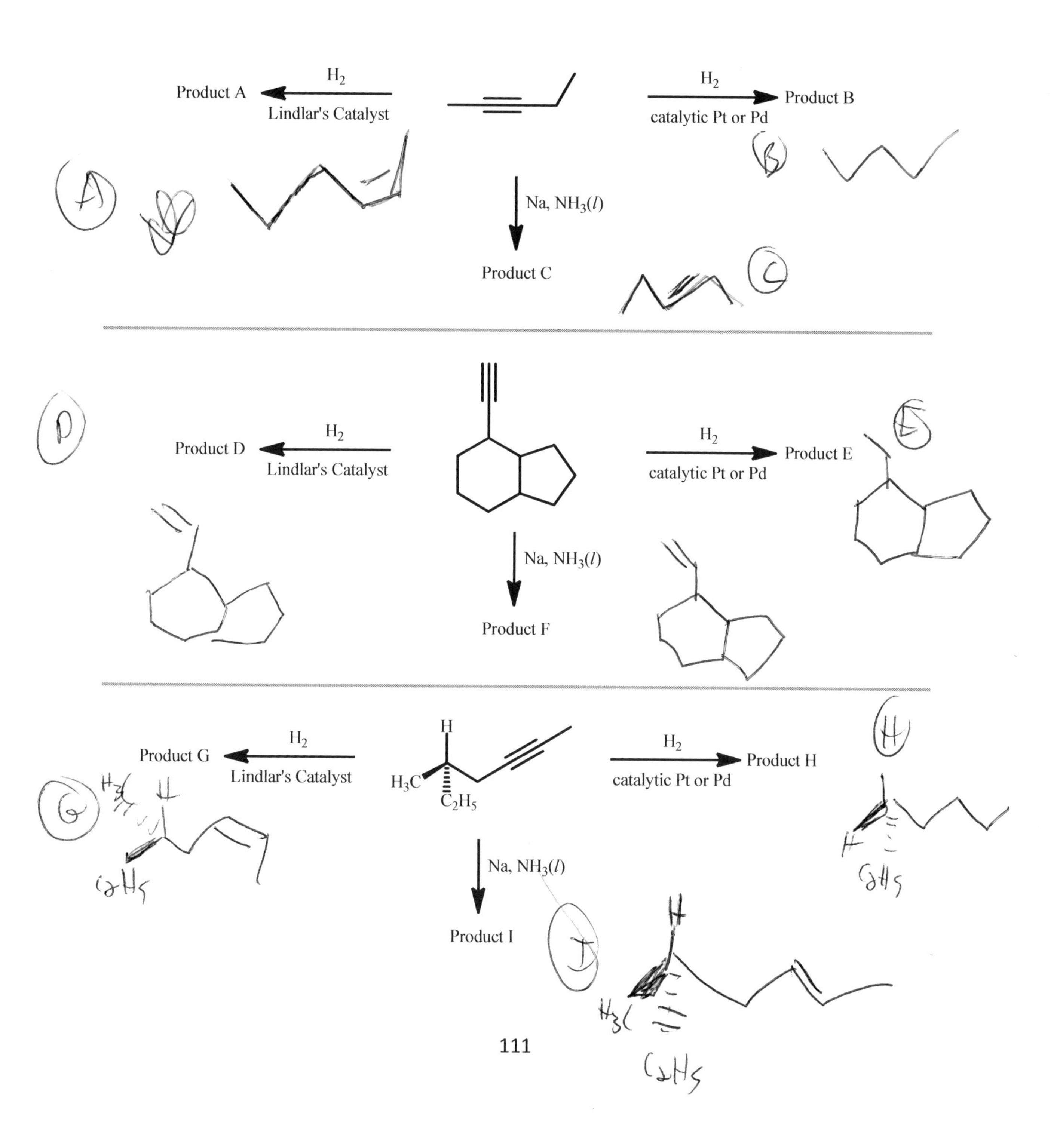

g) Answers to Self Test

i) Major product(s):

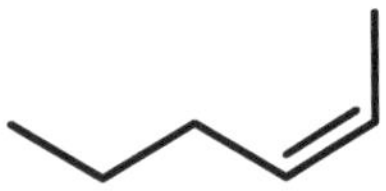

Product A

Product B

Product C

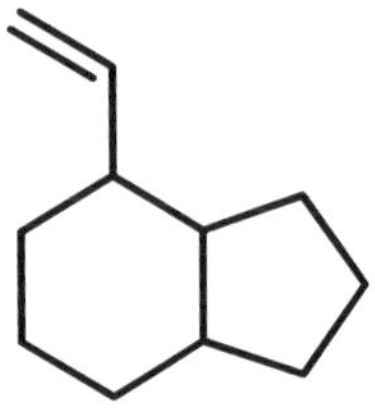

Product D

Product E

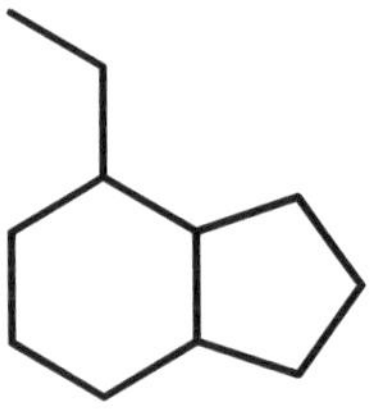

Product F

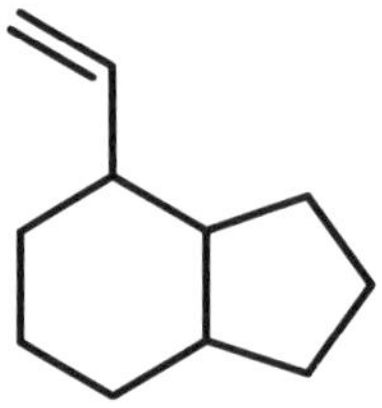

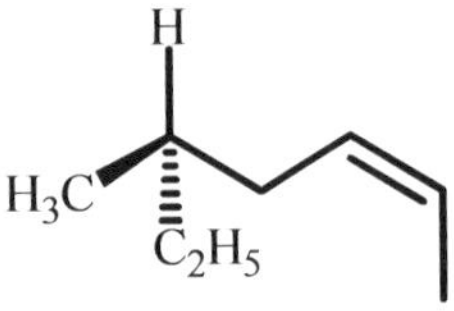

H

H_3C

C_2H_5

Product G

Product H

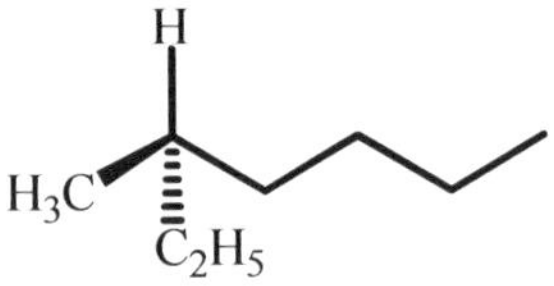

H

H_3C

C_2H_5

Product I

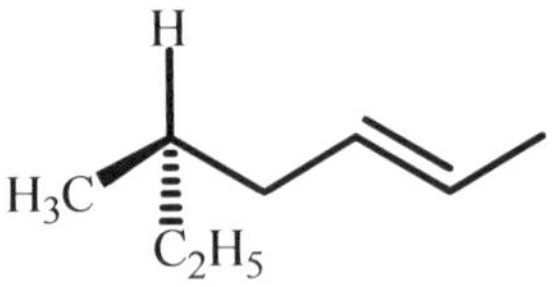

H

H_3C

C_2H_5

8. Use of acetylide anion as a nucleophile

a) General Net Reaction:

$$R-C\equiv C-H \xrightarrow[\text{2. R'-LG}]{\text{1. }NaNH_2} R-C\equiv C-R'$$

LG = leaving group (a relatively stable anion)

b) Notable Points

- The H atom of a terminal alkyne can be removed by a strong base such as $NaNH_2$. Alkyne **must be terminal** or there is no H to take off of the *sp*-hybridized C!
- The deprotonated alkyne can act as a nucleophile in reactions such as the S_N2 reaction (See Section I, reaction 5 and Section IV, reaction 2)
- When the acetylide derivative anion undergoes S_N2 reaction with an alkyl halide or tosylalkane, an internal alkyne is produced
- If the center from which the leaving group (LG) is a chiral center, inversion of configuration occurs, just as with other S_N2 reactions

c) Section Covered in Text:

d) Date(s) Discussed in Class

e) Arrow-Pushing Mechanism

1. preparation of the acetylide anion

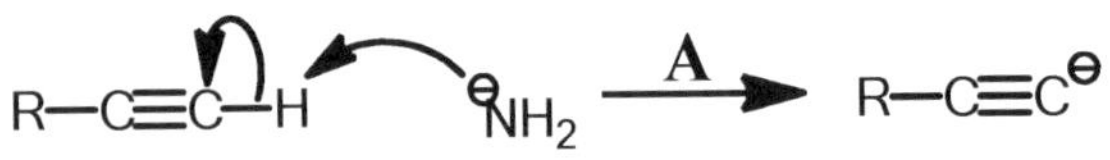

$$R{-}C{\equiv}C{-}H \quad {}^{\ominus}NH_2 \xrightarrow{\textbf{A}} R{-}C{\equiv}C^{\ominus}$$

2. Use of the acetylide anion as a nucleophile

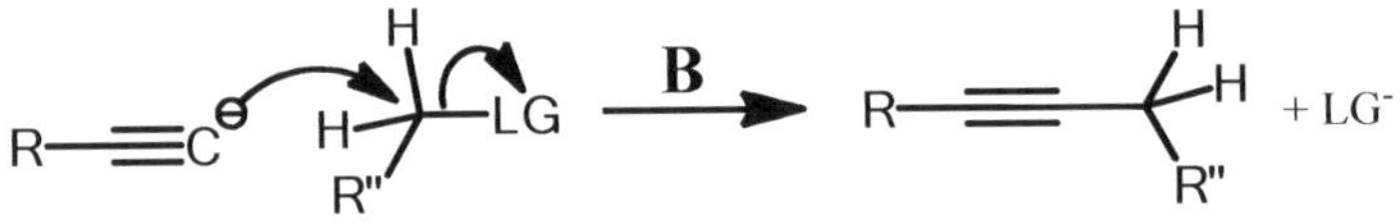

$$R{-}C{\equiv}C^{\ominus} \quad H_2C(R'')\text{–}LG \xrightarrow{\textbf{B}} R{-}C{\equiv}C{-}CH_2R'' + LG^{-}$$

LG = "leaving group"

f) Self Test

i) Draw the major product(s) of each reaction:

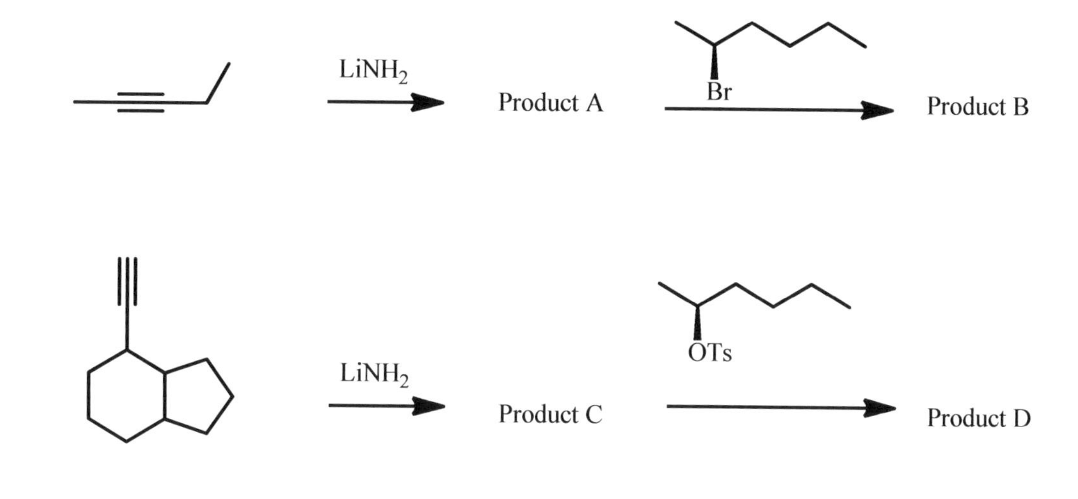

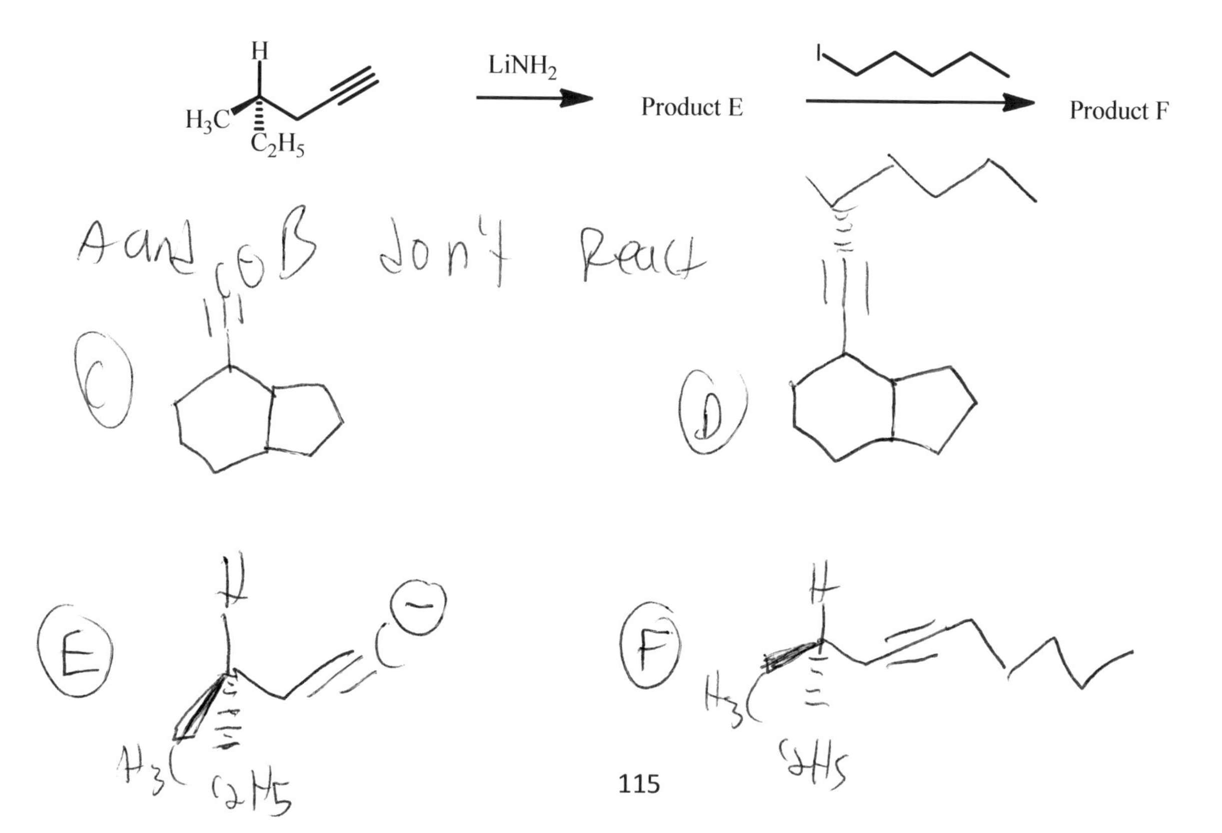

g) Answers to Self Test

i) Major product(s):

No Reaction; the alkyne is not terminal and cannot be deprotonated

Product A

Product B

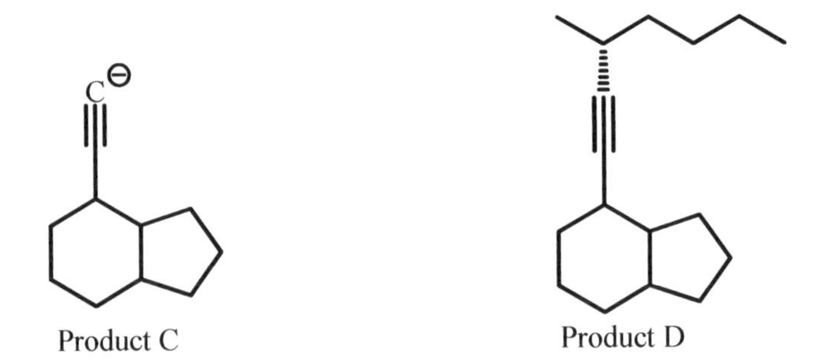

Product C

Product D

H

H_3C

C_2H_5

C

Product E

H

H_3C

C_2H_5

Product F

IV. Substitution and Elimination Reactions

Note: Because these reactions can compete with one another, a combined Self Test for all the reactions in this section is provided at the end of the section rather than having an individual Self Test at the end of each reaction.

1. The S_N1 Reaction

a) General Net Reaction:

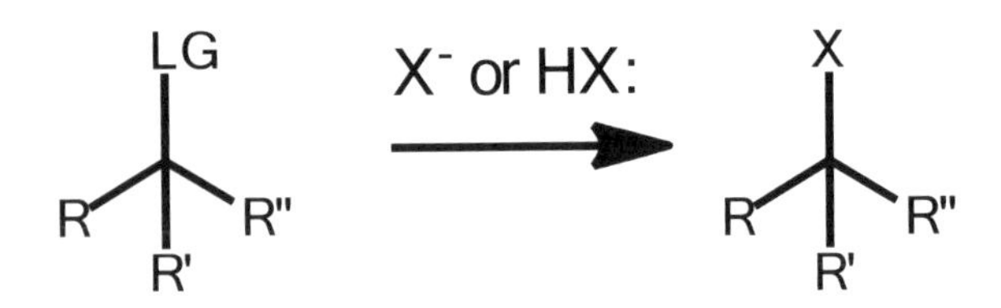

b) Notable Points

- Polar protic solvents are the best for this reaction
- If the leaving group (LG) is on a chiral center, there is net racemization of the center, leading to a 1:1 mixture of *R* and *S* configuration at the site
- The order from most to least reactive is: 3° > 2° > 1° center from which the leaving group leaves
- A carbocation intermediate is involved and can rearrange
- Substrates that have a LG on a 1° center do not generally react via this mechanism because the 1° carbocations are not very stable
- The E1 reaction often competes with this reaction because both mechanisms involve the same type of carbocation intermediate

c) Section Covered in Text:

d) Date(s) Discussed in Class

e) Arrow-Pushing Mechanism

i) General Reaction

LG

H_3C CH_3 CH_3

A

-LG-

H_3C ⊕ CH_3 CH_3

$X^{\ominus}$

B

CH_3 H_3C CH_3

X

ii) Stereochemistry Note

Note that if the leaving group leaves a chiral center, it does not matter whether you start with the (*R*)- or (*S*)-isomer, the carbocation intermediate produces is the same:

(*R*)-isomer
LG
C_2H_5
H
CH_3
(*S*)-isomer
LG
H_3C
H
C_2H_5
C_2H_5
H
CH_3

The carbocation is not chiral, so when the carbocation reacts with an achiral nucleophile the product cannot be one specific enantiomer. There is an equal probability that the nucleophile will attack the 'top' lobe of the empty *p* orbital (left option in the graphic below) or the 'bottom' lobe of the empty *p* orbital (right option in the graphic below):

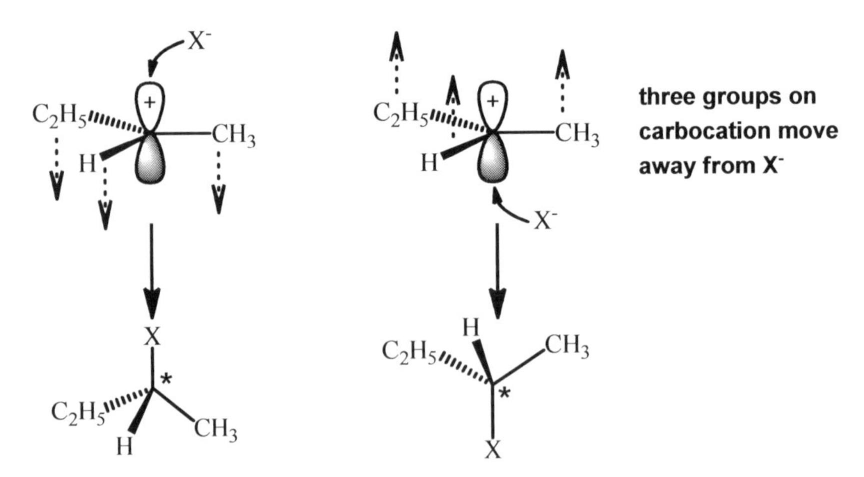

The result is that the product is an equal mixture of two enantiomers (a **racemic mixture**) even if the starting material is a pure enantiomer with a leaving group on the chiral center.

2. The S_N2 Reaction

a) General Net Reaction:

$$\text{R(LG)(R')CH} \xrightarrow{Nu^- \text{ or } HNu:} \text{R(Nu)(R')CH} + LG^- \text{ or } HLG$$

b) Notable Points

- Polar aprotic solvents are the best for this reaction, and more polar is better
- If the leaving group (LG) is on a chiral center, there is inversion of the center's configuration (called Walden Inversion)
- The order from most to least reactive is: 1° > 2° > 3° center from which the leaving group leaves
- No carbocation or other intermediate is involved; the reaction is concerted
- A good nucleophile is required for this reaction
- If the nucleophile is also a strong base, the E2 reaction will often compete with the S_N2 reaction

c) Section Covered in Text:

d) Date(s) Discussed in Class

e) Arrow-Pushing Mechanism

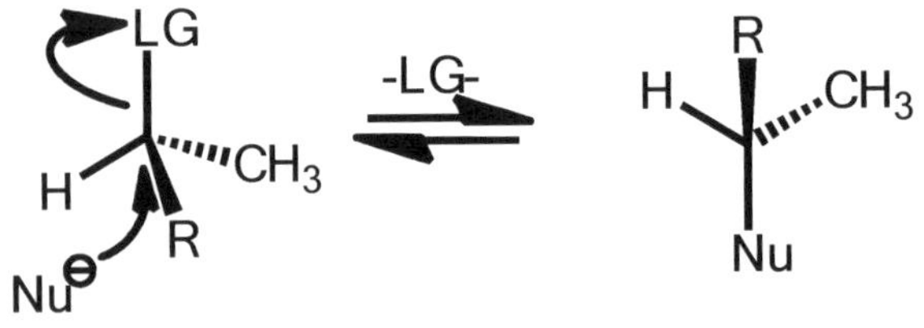

3. The E1 Reaction

a) General Net Reaction:

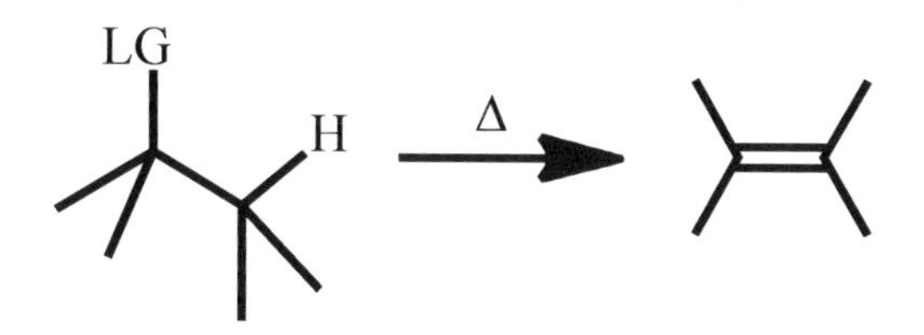

b) Notable Points

- Polar protic solvents are the best for this reaction.
- The order from most to least reactive is: 3° > 2° > 1° center from which the leaving group leaves
- A carbocation intermediate is involved and can rearrange
- Substrates that have a LG on a 1° center do not generally react via this mechanism because the 1° carbocations are not very stable
- If there is a choice, the more stable alkene is favored (Zaitsev's rule)
- The S_N1 reaction often competes with this reaction because both mechanisms involve the same type of carbocation intermediate

c) Section Covered in Text:

d) Date(s) Discussed in Class

e) Arrow-Pushing Mechanism

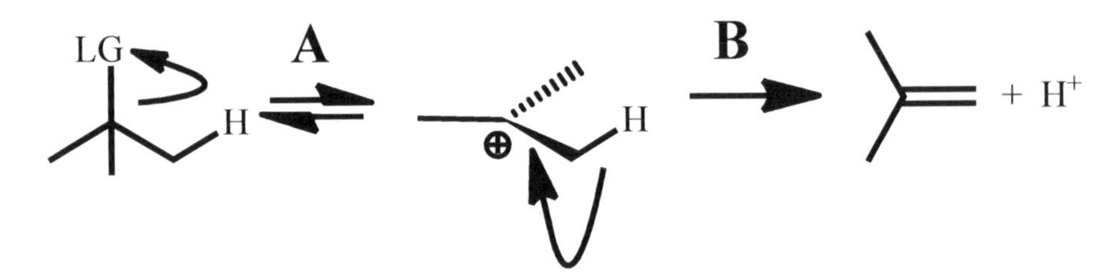

4. The E2 Reaction

a) General Net Reaction:

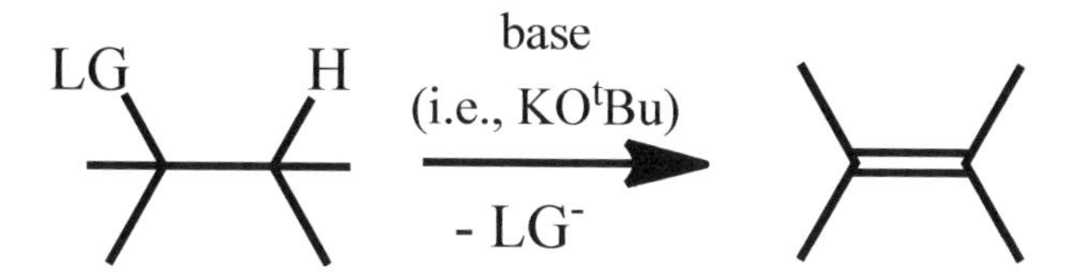

b) Notable Points

- Polar aprotic solvents are the best for this reaction, and more polar is better
- The leaving group (LG) and the H that leaves must be antiperiplanar to one another (see note in part e)
- If there is a choice, the more stable alkene is favored (Zaitsev's rule)
- The order from most to least reactive is: 3° > 2° > 1° center from which the leaving group leaves
- No carbocation or other intermediate is involved; the reaction is concerted
- A strong base is required for this reaction
- If the base is also a good nucleophile, the S_N2 reaction will often compete with the E2 reaction

c) Section Covered in Text:

d) Date(s) Discussed in Class

e) Arrow-Pushing Mechanism

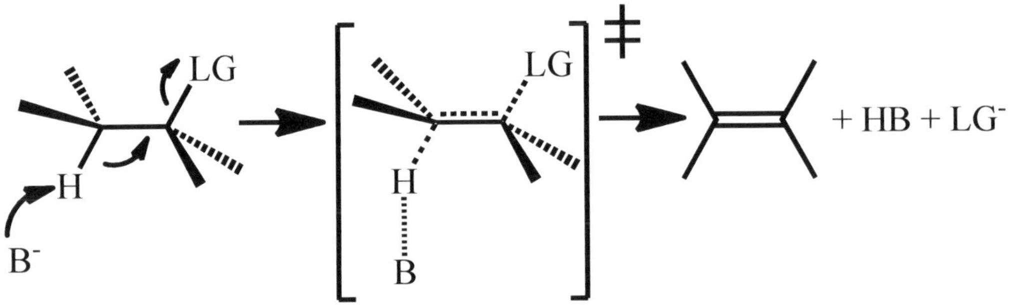

LG
H
B⁻
LG
H
B
‡
+ HB + LG⁻

NOTE:

The H and the LG must be *antiperiplanar* to one another (i.e., they would appear anti to one another in the Newman projection). Because the reaction is concerted, the other groups on the starting material do not rearrange during reaction.

5. Self Test and Answers for S_N1, S_N2, E1 and E2 Reaction

a) Self Test

i) Choosing from the four solvents shown, which is the best choice for:

Solvent choices

diethyl ether | dimethylsufoxide (DMSO) | *n*-propanol | *n*-hexane

S_N1 reaction?

S_N2 reaction?

E1 Reaction?

E2 Reaction?

ii) Choosing from the four substrates shown, which will generally react fastest by:

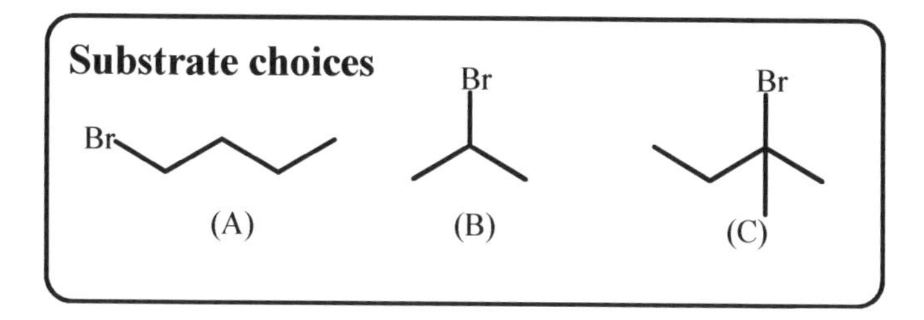

S_N1 reaction?

S_N2 reaction?

E1 Reaction?

E2 Reaction?

(continued on next page)

iii) For each reaction, circle the mechanism(s) by which the major products will form, and draw the major products. Be sure to indicate stereochemistry where relevant.

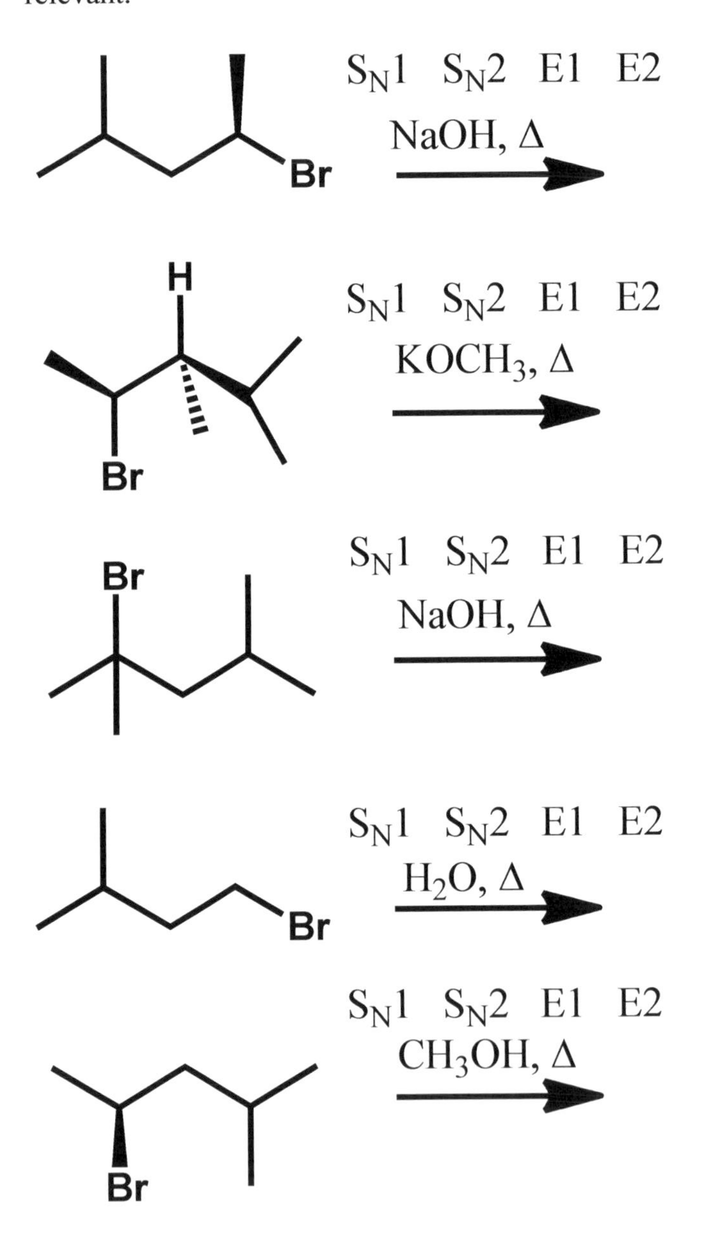

b) Answers to Self Test

i)

The best solvent for the S_N1 reaction is *n*-propanol

The best solvent for the S_N2 reaction is DMSO

The best solvent for the E1 Reaction is *n*-propanol

The best solvent for the E2 Reaction is DMSO

ii)

The fastest reaction via the S_N1 reaction is with substrate C

The fastest reaction via the S_N2 reaction is with substrate A

The fastest reaction via the E1 Reaction is with substrate C

The fastest reaction via the E2 Reaction is with substrate A

(continued on next page)

iii)

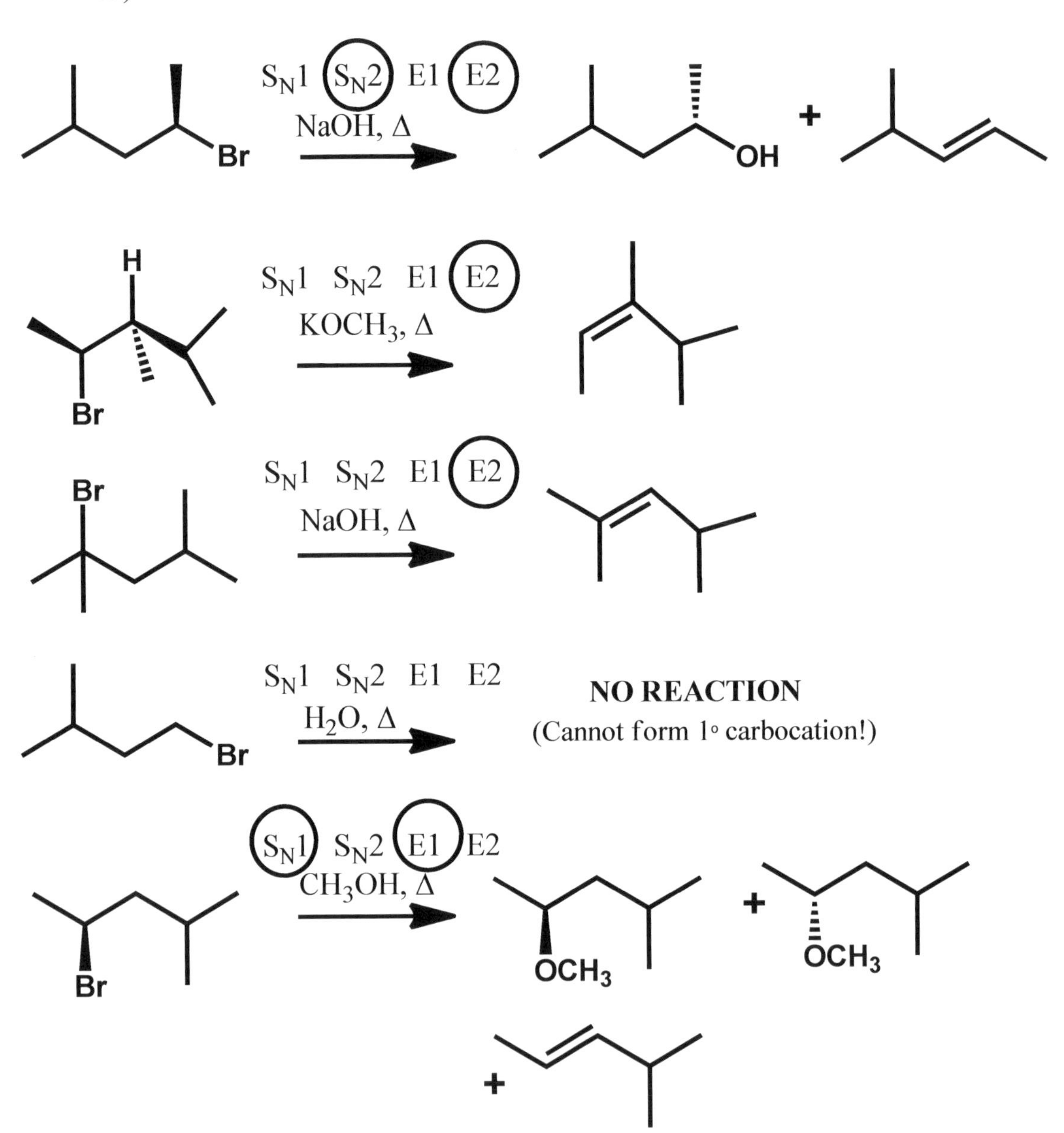

Br
S_N1 S_N2 E1 E2
NaOH, Δ
OH
+
H
Br
$KOCH_3$, Δ
Br
NaOH, Δ
Br
H_2O, Δ
NO REACTION
(Cannot form 1° carbocation!)
Br
CH_3OH, Δ
OCH_3
OCH_3
+

Made in the USA
Lexington, KY
31 October 2011